Ocean Modelling for Beginners

Jochen Kämpf

Ocean Modelling for Beginners

Using Open-Source Software

Springer

Assoc. Prof. Jochen Kämpf
School of Environment
Flinders University
Chemistry and Earth Sciences
P.O. Box 2100
Adelaide SA 5001
Australia
jochen.kaempf@flinders.edu.au

Additional material to this book can be downloaded from http://extra.springer.com

ISBN 978-3-662-50233-4 ISBN 978-3-642-00820-7 (eBook)
DOI 10.1007/978-3-642-00820-7
Springer Heidelberg Dordrecht London New York

Cover design: Bauer, Thomas

Printed on acid-free paper

Springer is part of Springer Science+Business Media (www.springer.com)

Preface

This book focuses on the dynamics of the ocean being influenced by the Earth's rotation and density stratification. Fluids in motion are a difficult subject of study that traditionally requires advanced knowledge of analytical mathematics, in particularly matrix algebra, differential and integral calculus, and complex analysis. Hence, this fascinating field of science, known as *geophysical fluid dynamics*, is accessible only to a limited number of students – those who either are naturally geniuses or those who underwent tough years of intense University study.

Fluid processes are inherently complex and analytical solutions describing fluid dynamics exist only in a few instances and only under highly simplified assumptions. Computer-based numerical models are required to approximate fluid behavior in more realistic situations. Because of its complexity, universities tend to offer subjects in computational modelling of fluid dynamics only at postgraduate level. This is a pity given that fluid processes are truly fascinating in nature and given that the oceans play a significant role in shaping life on Earth.

The approach I pursue in this book is different from the traditional approach. Here, numerical models are gradually built up and refined with the objective to illustrate and explore various dynamical processes occurring in fluids. Little mathematical background knowledge is required, and the focus is placed where it should be, namely on the physics inherent with fluids in motion. This book is a combination of a textbook and a workbook including more than 20 computer-based exercises, written in FORTRAN 95. Analytical solutions of certain fluid phenomena are used as invaluable benchmarks for verification of these model simulations. In parallel to this book, the reader is encouraged to consult textbooks by Cushman-Roisin (1994), Pond and Pickard (1983) and Gill (1982).

The modelling-based approach has many advantages over the traditional analytical approach and, in the author's belief, will open the field of geophysical fluid dynamics to a much broader audience. Obvious advantages are that (a) complex fluid processes such as barotropic or baroclinic instabilities, otherwise exclusively reserved to experts, can be studied by a lay person, (b) instead of still pictures of results, the reader can create animations of processes, and (c) the reader can adopt computer codes, provided in this book, in a modified form for own independent studies. Without doubt, learning is greatly enhanced by playing and this book provides the reader with the tools (or toys) to achieve this.

Access to a standard computer is the only requirement for the completion of exercises. All computer software suites required are open-source programs being freely available for download from the Internet. This book is designed such as to keep financial burden for the interested reader at a minimum. Background knowledge in scientific computing is an advantage but not a requirement.

This book introduces the reader to conservation principles obeyed by fluids in motion, the finite-difference formulation of these principles, and provides the reader with a step-to-step guide to so-called *finite-difference layer modelling*. This book details numerical methods including a flooding algorithm, semi-implicit treatments of both the Coriolis force and bottom friction, and total-variation diminishing (TVD) advection schemes that are absolute minimum requirements for adequate modelling of fluid processes. Further simplification seems not possible, but there are certainly more accurate (but also mathematically more difficult) methods available. A description of higher-order, more complex methods is beyond the scope of this book.

I dedicate this book to my doctorfather Professor Jan O. Backhaus for his creativity and overwhelming enthusiasm which have been the prime motivation for me to pursue a career in the field of physical oceanography. Many of Jan's suggestions and approaches to numerical modelling are implemented in this book.

Other invaluable sources of motivation behind this work are the classical books of Henry Stommel, namely "An Introduction to the Coriolis Force" published in 1989 and co-authored by Dennis Moore, and "A View of the Sea", published in 1987. Similar to the approach I take here, Stommel's work underpinned theory with computer programs, written in BASIC, that can be run by the reader for independent studies.

Adelaide, Australia,
January 2009 J. Kämpf

Contents

Chapter 1
Requirements

Abstract This chapter lists the software packages required including download links. FORTRAN 95 will be employed as the main programming language. SciLab will be used for graphical display and animations of results. Some useful tips are given on how the reader should organise his/her work.

1.1 Software Overview

The following classes of software are required for this book:

- a suitable programming language including compiler to create executable programs of your codes
- an advanced scientific data visualisation program to produce graphs and animations
- a text editor to write model codes.

There are many excellent professional software packages on the market, but they typically come at a fairly high cost. Luckily, there are so-called *open-source* programs, being free of charge and running on most computer platforms. Exercises of this book are exclusively based on such open-source software. Hence, apart from having access to a personal computer, there are no additional financial burdens that could prevent the reader from becoming an experienced fluid modeller!

1.2 Programming Language and Compiler

Exercises of this book use the FORTRAN 95 programming language. Although this language is relatively old (the first version was produced in 1957), it is still commonly used by the scientific community, in particular for large model applications run on the world's biggest computer platforms, so-called Supercomputers.

Model codes of this book have been tested with the FORTRAN compiler G95 (Version 0.91) that can be downloaded from:

http://www.g95.org/

J. Kämpf, *Ocean Modelling for Beginners*,
DOI 10.1007/978-3-642-00820-7_1, © Springer-Verlag Berlin Heidelberg 2009

Microsoft Windows Operation systems, for instance, this compiler is accessible from the Command Window, found under Start/All programs/Accessories. The compiler is called by entering the command:

g95 filename.f95 <enter>

If free of errors, this produces an executable file of standard name "a.exe", that, once executed, does hopefully the tasks it was designed for. Consult the G95 user manual (available from the above Web page) for more advanced compiler options. Enter "dir" in the Command Prompt window for a list of all subfolders and files contained in a folder. Enter "cd *name*" to move to another subdirectory. Use "cd .." to move back one level.

1.3 Data-Visualisation Software

To produce graphs and animations of model outputs, I can highly recommend the open-source software SciLab that can be downloaded from:

http://www.scilab.org/

I have used Version 4.1 for Windows. This software is also available for GNU/ Linux platforms. The good thing about this software is that it comes for free and that its commands resemble those in MATLAB, being used by many scientists. SciLab also comes with a text editor that the reader can use to write FORTRAN codes.

Some of the SciLab commands used in the scripts of this book are obsolete in the newest SciLab 5.x versions. Updated scripts can be requested from the author per email.

SciLab allows for export of graphics to several formats including POSTSCRIPT and GIF. The software suite ImageMagick, downloadable at:

http://www.imagemagick.org/

can be used to read, convert and write images in a variety of formats (about 100) including GIF, JPEG, JPEG-2000, PDF, PhotoCD, PNG, Postscript, and TIFF. ImageMagick can also be used to produce GIF animations that can be watched on many internet browsers. An option is the open-source Firefox Web browser developed by MOZILLA found at:

http://www.mozilla.org/

1.4 Text Editor

Any basic text editor such as Notepad or Wordpad will do for the writing of FORTRAN programs. The open-source editor VIM can be downloaded at:

http://www.vim.org/

SciLab has also a text editor that can also be used to write FORTRAN codes.

1.5 Organisation of Work

After installation of all software suites required, the reader should create a folder called "EXERCISES" as an archive for all exercises that follow. During the progress of work, a number of subfolders should be created within this folder each of which refers to a certain exercise. All files belonging to a certain exercise (such as FORTRAN 95 codes and SciLab scripts) should be kept in a single subfolder.

1.6 Structure of this Book

The book describes a variety of dynamical processes occurring in this ocean accompanied by practical exercises. During the course of progress, the level of complexity of processes will increase and so will the length of numerical codes. Consequently, the reader should study exercises in consecutive order. In order to improve the learning outcome, I strongly recommend that the reader attempts to write own model codes from scratch. If this leads to unsatisfactory results, the reader is invited to employ numerical codes that are supplied on the CD-ROM to this book.

Chapter 2
Motivation

Abstract This chapter introduces the reader to the "world" of numerical modelling using the decay problem as a first benchmark. Discussed are finite differences, explicit and implicit schemes, and conditions of consistency, accuracy, stability, and efficiency. FORTRAN codes and SciLab scripts are used to create a first numerical prediction model and graphical display of results.

2.1 The Decay Problem

2.1.1 The Problem

The so-called *decay problem* is chosen as a first example to demonstrate what numerical modelling is. In mathematical terms, this problem can be expressed as:

$$\frac{dC}{dt} = -\kappa \cdot C \tag{2.1}$$

where C is concentration of a substance, t is time, and κ (the Greek symbol "kappa") is a positive constant parameter. The d symbol refers to a change of a variable with respect to some other parameter such as time as in the above equation.

2.1.2 Physical Interpretation

The term on the left-hand side of Eq. (2.1) refers to the temporal change of concentration per time unit. The right-hand side specifies this temporal change. For $\kappa = 0$, there will no change and C remains unchanged at its initial value. With $\kappa \neq 0$, on the other hand, the right-hand side is negative since concentration is always a positive quantity. Accordingly, C will gradually decrease with time at a rate in proportion to concentration itself at any time instance.

J. Kämpf, *Ocean Modelling for Beginners*,
DOI 10.1007/978-3-642-00820-7_2, © Springer-Verlag Berlin Heidelberg 2009

In mathematics, Eq. (2.1) is termed a *first-order ordinary differential equation* and, as it requires an initial value for C, it is also referred to as an *initial-value problem*.

2.1.3 Example

Grandma used to keep a 10-litre carton of red wine in a storage room. I could not resist and crept every night to this room to get myself 1 litre of wine. To avoid grandma noticing the fast disappearance of the wine, each night, I topped up the carton with tab water. However, I was caught soon or do you really think that grandma would not notice changes in the wine's taste and color owing to dilution with water? Anyway, the reader can easily work out how the concentration of wine changed with every day of my early-year drinking habit. All that needs to be done is to take away 10% of the wine concentration on a daily basis. The following table shows the result of this.

Day	Wine content (l)	Wine concentration (%)
0	10	100
1	9.0	90.0
2	8.1	81.0
3	7.29	72.9
4	6.56	65.6
5	5.9	59.0
6	5.31	53.1
7	4.78	47.8
8	4.3	43.0
9	3.87	38.7
10	3.49	34.9
11	3.14	31.4
12	2.82	28.2
13	2.54	25.4
14	2.29	22.9
15	2.06	20.6
16	1.85	18.5
17	1.67	16.7
18	1.5	15.0
19	1.35	13.5
20	1.22	12.2
21	1.09	10.9
22	0.98	9.8

Figure 2.1 shows a graph of the outcome of this first numerical prediction model, only using a piece of paper and a pen. As expected, there is a decrease of wine content as the days go past. In fact, this decrease is approximately exponential.

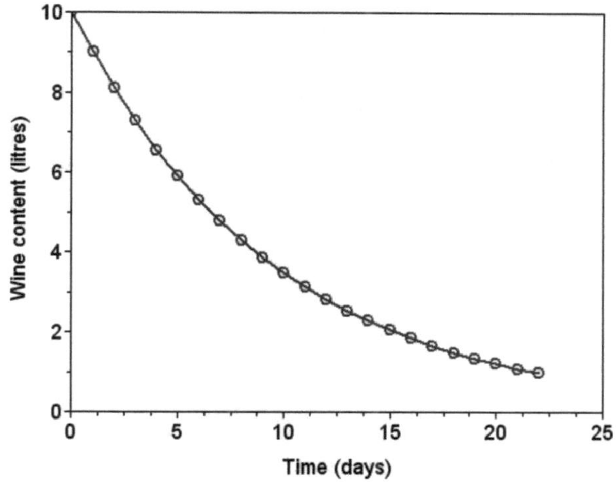

Fig. 2.1 Evolution of wine content

2.1.4 How to Produce a Simple Graph with SciLab

This exercise explains how to use SciLab to produce Fig. 2.1.

Step1: Open a suitable text editor and insert and save the three rows of the above table (without header) in a file called, for instance, "winethief.txt".

Step 2: Fire up SciLab and change to the folder containing this file.

Step 3: Enter the following commands (each followed by <return>):

```
x=read("winethief.txt",-1,3);
plot(x(:,1),x(:,2));
xtitle(" ","Time (days)", "Wine content (litres)"," ")
```

The "read" command imports data from the file. The parameter "−1" is used if the number of rows is unknown. "3" means: read the first three columns, "−1" does not work for columns. The above plot command plots values of row 2 against those of row 1. The last line adds axis labels. The semicolon ";" suppresses output to the SciLab command window.

Step 4: Change line width, font size, etc. in the graphics window by selecting "Edit Figure Properties" in the "File" menu or by clicking "GED" in the graphics window. The reader is encouraged to "play around" with all options available.

Step 5: Export the graph using a suitable format. I often selected the PostScript format and converted images into Portable Network Graphics files (PNG) with ImageMagick.

2.2 First Steps with Finite Differences

2.2.1 Finite Time Step and Time Level

With the use of a discrete *time step* Δt, we may formulate (2.1) as:

$$\frac{C^{n+1} - C^n}{\Delta t} = -\kappa C^n \tag{2.2}$$

where the integer n refers to a certain *time level*. This time index must not be confused with "to the power of". Conventionally, $n = 0$ gives the concentration at start time of your simulation, $n = 1$ refers to the concentration after one time step (of Δt in duration), $n = 2$ refers to the concentration after two time steps, and so on.

2.2.2 Explicit Time-Forward Iteration

It is convenient to move the unknown variable in (2.2) to the left-hand side of the equation and shuffle all known terms to the right-hand side. This gives:

$$C^{n+1} = C^n - \Delta t \cdot \kappa \cdot C^n = (1 - \Delta t \cdot \kappa)\, C^n \tag{2.3}$$

where $C^{n=0}$ refers to the initial concentration that needs to be prescribed together with values of κ and Δt. This iterative method uses values known at a certain time level n to predict the value of C at the next time level $n + 1$ and is therefore called *explicit time-forward iteration*.

2.2.3 Condition of Numerical Stability for Explicit Scheme

As can be seen from (2.3), with every time step, the concentration becomes decimated by a certain fraction in an iterative manner. This fraction is given by the product $\kappa \cdot \Delta t$. It is at hand to request that this product be less than unity, otherwise the predicted concentration would become negative, which would not make sense. For $\kappa \cdot \Delta t > 2$, the magnitude of concentration would even increase. The corresponding condition:

$$\Delta t < \frac{1}{\kappa} \tag{2.4}$$

is called a *condition of numerical stability*. Hence, the prediction of (2.3) is only stable when (2.4) is satisfied. Accordingly, the maximum time step that can be chosen depends on the value of κ.

2.2.4 Implicit Time-Forward Iteration

Alternatively, Eq. (2.1) can be formulated in finite-difference form as:

$$\frac{C^{n+1} - C^n}{\Delta t} = -\kappa \cdot C^{n+1} \tag{2.5}$$

where the concentration on the right-hand side is evaluated at the next time level $n + 1$. This approach might sound strange to some readers, but if we reorganize this equation, we yield a clear separation of known and unknown terms of the form:

$$C^{n+1} = \frac{C^n}{(1 + \Delta t \cdot \kappa)} \tag{2.6}$$

The clear advantage of this *implicit scheme* over the explicit approach is that it is numerically stable for any value of Δt. The denominator in the later equation always exceeds unity, so that concentration gradually decreases with time (and never changes sign).

2.2.5 Hybrid Schemes

One could also use a mix between the explicit and the implicit scheme, which can be formulated as:

$$\frac{C^{n+1} - C^n}{\Delta t} = -\alpha \cdot \kappa \cdot C^{n+1} - (1 - \alpha)\kappa \cdot C^n \tag{2.7}$$

where the weighting factor α (the Greek symbol "alpha") has to be chosen from a range between zero and unity. The choice of $\alpha = 1$ gives the fully implicit scheme, whereas $\alpha = 0$ leads to the fully explicit scheme. With $\alpha = 0.5$, we obtain a so-called *semi-implicit* scheme.

2.2.6 Other Schemes

There are more advanced schemes such as the "Runge-Kutta scheme" or the "Adams-Bashforth scheme", not discussed here, that in addition to current and future time level consider a number of sub-time steps. The accuracy and efficiency of the prediction model can be significantly improved with such schemes.

2.2.7 Condition of Consistency

The exact analytical solution of the decay problem (2.1) for an initial concentration of C_o is given by:

$$C(t) = C_o \exp(-\kappa \cdot t) \tag{2.8}$$

where "exp" is the exponential function. A numerical model is said to be *consistent* if its finite-difference solution converges toward the solution of the governing differential equation when the numerical time step (or grid size) is made vanishingly small. This implies that the concentration predicted by our model should get the closer to the true solution for a decrease of the time step Δt.

2.2.8 Condition of Accuracy

A certain error referred to as *truncation error* is made when using finite differences. *Round-off errors* are another source of error, being related to the fact that computers can represent numbers only with a finite number of digits. Both errors should stay reasonably small over the duration of a simulation.

2.2.9 Condition of Efficiency

Large programs may require substantial computer space for data output and storage, and completion of model runs may take a long time. Hence, model codes have to be written in an efficient manner such that the task is completed within a reasonable time span and without "stuffing up" the computer with enormous amounts of data.

2.2.10 How Model Codes Work

The compiler translates the FORTRAN 95 code line by line and *from top to bottom*. This implies that parameters must be declared and specified before they can be manipulated. Declaration means specification of the type of the parameter. There are integers, real numbers, arrays, characters and logical parameters.

Specification means allocation of values to parameters. In principle, each line of a computer code can only have a single unknown on the left-hand side of an equation, such as "$x = b + c$", where b and c have to be declared and assigned values farther up in the code, and x has to be at least declared.

2.2.11 The First FORTRAN Code

Write a first FORTRAN code that prints "Hello World" on the display. This is an old tradition among modellers. The solution is the code:

```
PROGRAM first
write(6,*)"Hello World"
END PROGRAM first
```

FORTRAN programs start with a PROGRAM *name* statement and finish with a END PROGRAM *name* statement. My program is called "first", but the reader is free to choose a different name. Save this file as "first.f95".

2.2.12 How to Compile and Run FORTRAN Codes

Open the Command Prompt window (on Windows systems this is found under Start => All Programs => Accessories => Command Prompt) and move to the folder containing your FORTRAN source file.

Step 1: Compile the program by entering the command:

```
g95 -o first.exe first.f95
```

where "-o" specifies the name of the executable program.

Step 2: Correct errors until the compiler does not return any error messages.

Step 3: If the compiling process was successful, the newly created file "first.exe" can be executed by simply double-clicking its icon in a file window or by entering "first" <return> in the Command Prompt window.

The result of this will be the display of "Hello World" in the Command Prompt window. Congratulations!

2.2.13 A Quick Start to FORTRAN

All constants, parameters, variables and arrays have to declared before use. Hereby, full numbers called "integer" ($-3, 0, 1, 3$, etc.) are distinguished from decimal numbers called "real" ($1.2, 4.2, -5.23$, etc.). Other options are logical expressions (true or false) and text (characters). Constant parameters are declared at the beginning of the code with:

```
INTEGER, PARAMETER :: nx = 11 ! horizontal dimension
INTEGER, PARAMETER :: nz = 5 ! vertical dimension
```

```
REAL, PARAMETER :: G = 9.81 ! acceleration due to gravity
REAL, PARAMETER :: RHOREF = 1028.0 ! reference density
REAL, PARAMETER :: PI = 3.14159265359 ! pi
```

Text after a pronunciation mark is treated as a comment and ignored by the compiler. Although not required, comments are very useful aids to highlight the structure of the program and as reminders for future uses. Parameters that are allowed to change values during the program's execution are declared with:

```
REAL :: wspeed ! wind speed
INTEGER :: k ! grid index
CHARACTER(3) :: txt
```

In this example, "txt" is a character with three letters. One-dimensional and two-dimensional arrays are declared with:

```
REAL :: eta(0:nx+1) ! sea-level elevation
REAL :: w(0:nz+1,0:nx+1) ! vertical velocity
```

With nx = 11 and nz = 5, for instance, "eta" obtains 11 + 2 = 13 so-far unassigned elements: eta(0), eta(1), $\cdots$ ..., eta(11), eta(12), and "w" is a two-dimensional array of 13 columns and 7 rows. After the declaration section, values can be assigned to these arrays using DO loops such as:

```
DO k = 0,nx+1
  IF(k > 50) THEN
    eta(k) = 1.0
  ELSE
   eta(k) = 0.0
  END IF
END DO
```

This DO-loop repeats certain calculations for the index "k" running from 0 at steps of 1 to nx+1. If the reader wants to do it the other way around, the solution is:

```
DO k = nx+1,0,−1
  IF(k > 50) THEN
    eta(k) = 1.0
  ELSE
    eta(k) = 0.0
  END IF
END DO
```

This example also includes an IF statement. Options are ">" (greater than), "<" (less than), "==" (equal to), ">=" (greater or equal), "<=" (less or equal), and "/="

(not equal). If there is only one line in an IF-statement, the "ELSE" and closing "END IF" statements can be dropped, such as in the following example:

```
DO k = nx+1,0,−1
  eta(k) = 0.0
  IF(k > 50) eta(k) = 1.0
END DO
```

Files for output can be opened with the statement:

```
OPEN(10, file = 'Ex1a.txt', form = 'formatted', status = 'unknown')
```

The first entry (10) is a reference unit number for subsequent WRITE or READ statements. The "file" entry specifies the desired filename. The "form" entry specifies whether to have ASCII numbers or binary numbers as output. I chose ASCII output. The status entry "unknown" implies new creation of a file if this does not exist, otherwise an existing file will be overwritten. Be careful not to overwrite files that might be needed in the future. Other status options are "new" or "old".

Output of data is done via "WRITE" statements such as

```
WRITE(10,*) G
```

where the unit number (10 here) refers to a file opened before, and the "*" symbol creates a standard format. Note that the unit number 6 is reserved for output to the screen as in our first FORTRAN code. Similarly, "READ(5,*)" reads input from the keyboard. Several outputs at once can be produced with:

```
WRITE(10,*) eta(10), eta(20), eta(30)
```

Doing this repeatedly, there will be three columns of data. Output of an entire row of an array is done with:

```
WRITE(10,*) (eta(k), k = 1,nx)
```

Files no longer needed for output should be closed with the statement:

```
CLOSE(10)
```

2.3 Exercise 1: The Decay Problem

2.3.1 Aim

The aim of this exercise is to predict the decay of a substance according to (2.1) using a FORTRAN code based on either the explicit or the implicit scheme.

2.3.2 Task Description

Consider a substance that has an initial concentration of 100% and use a decay constant of $\kappa = 0.0001$ per second (or $\kappa = 10^{-4}\ \mathrm{s}^{-1}$). Choose different values of the time step to verify whether the prediction becomes more accurate for a finer temporal resolution. Explore both the explicit and the implicit scheme.

2.3.3 Instructions

Use any text editor to write the FORTRAN code and save the file under the name "Exercise1.f95". Blanks or other unusual symbols are not permitted here. Other filenames may be used, but the reader should make sure that the filename is not too long and that it has something to do with the exercise. The file extension "f95" identifies this file as a FORTRAN 95 source code.

2.3.4 Sample Code

The Fortran code for this exercise, called "winethief.f95" can be found in the folder "Exercise 1" on the CD-ROM accompanying this book.

2.3.5 Results

As a result of the model run, the data output files "output1.txt" or "output2.txt" should appear in the file list. The MODE parameter in the code switches between the explicit and the implicit schemes. To avoid the recompiling procedure, values for "mode" could be alternatively read from the keyboard with "READ(5,*) mode".

Figure 2.2 shows model results for a time step of $\Delta t = 3600\,\mathrm{s}$ using either the explicit scheme (2.3) or the implicit scheme (2.6). As can be seen, the explicit scheme slightly underestimates the correct concentration, whereas the implicit scheme slightly overestimates concentration. A semi-implicit approach would probably give the best solution, but this remains to be verified by the reader.

With a time step of 3600 s, completion of the model run took only a few seconds on my computer. The accuracy of the prediction can be substantially improved with choice of a much finer temporal resolution with a time step of, say, $\Delta t = 1\,\mathrm{s}$, which the reader can easily verify.

2.3.6 Additional Exercise for the Reader

Repeat this exercise with use of the hybrid scheme (2.7) and explore the solutions for $\alpha = 0.25$, 0.5 and 0.75.

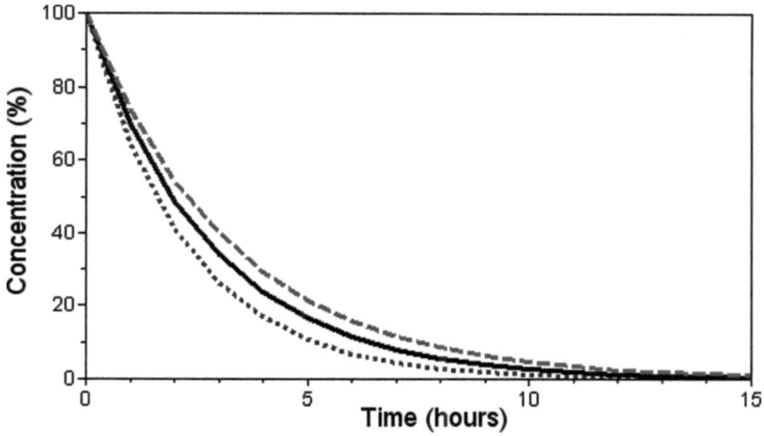

Fig. 2.2 Evolution of concentration (%) as a function of time. The *upper* and *lower curves* show results using the implicit and the explicit schemes, respectively. The *middle curve* displays the analytical solution according to Eq. 2.8

2.4 Detection and Elimination of Errors

2.4.1 Error Messages

If a FORTRAN code contains errors, the compiler will return one or more error messages. There are a few important steps to follow for successful detection and elimination of errors.

2.4.2 Correct Errors One by One

Only correct one error at a time with reference to the first error message. Often other errors are just followers of the first one. Similarly, an important rule is that the code should be compiled after each single alteration made. It can be tedious to locate errors after a dozen changes have been implemented without verification. Errors are often the result of a lack of concentration.

2.4.3 Ignore Error Message Text

Occasionally, the compiler's error messages are confusing and misleading. Therefore, I recommend to ignore message text and rather focus on the line number this message refers to.

2.4.4 Frequent Errors

Errors are frequently associated with misspelling such as “0” (zero) instead of the letter “o” or vice versa, or mistaking “1” (one) as the letter “l”. Often an ENDDO, ENDIF or a bracket is just missing.

2.4.5 Trust Your Compiler

Important is to always trust the compiler. If there is an error message then there is an error in your code. The error message text might be misleading, but the presence of at least one error is a fact. In a state of frustration after endless unsuccessful search after errors, the reader should take a good rest, perhaps a walk or a sleep. Breaks are always important!

2.4.6 Display Warnings

Warnings can be displayed with addition of “-Wall” (display all warnings) in the compiling command. For Exercise 1, for example, this command reads:

```
g95 -Wall -o winethief.exe winethief.f95
```

Warning messages should be explored for potential errors in the code.

Chapter 3
Basics of Geophysical Fluid Dynamics

Abstract This chapter introduces the reader to scalars and vectors, Newton's laws of motion, waves and steady-state flows, the buoyancy force, the Coriolis force, a number of conservation principles, turbulence, and, finally, to the Navier–Stokes equations of fluid motion. Practical exercises focus on the interference of waves, movement of a buoyant object in a stratified water column, and movement of objects under the sole influence of the Coriolis force.

3.1 Units

The International System of Units will be used in most instances. Distances are expressed in metres (m), time lapse in seconds (s), and mass in kilograms (kg). For convenience, temperatures are expressed in degrees Celsius (°C), which is the thermodynamic temperature in Kelvin (K) plus a constant of 273.16. Time lapse is translated into more convenient units such as minutes, hours, days, weeks, months, years, etc. Other units can be derived from the above base units. For example, volume is expressed in cubic metres (m^3). Often, we also use symbols for multipliers such as:

$$\text{micro } (\mu) = 10^{-6} = 0.00001$$
$$\text{milli (m)} = 10^{-3} = 0.001$$
$$\text{centi (c)} = 10^{-2} = 0.01$$
$$\text{kilo (k)} = 10^{3} = 1000$$

3.2 Scalars and Vectors

3.2.1 Difference Between Scalars and Vectors

A *scalar* is a physical quantity without directional information. Temperature and pressure are examples of this. A *vector*, on the other hand, carries information of both magnitude and direction. Velocity is an example of this. A car moves at a certain speed into a certain direction.

J. Kämpf, *Ocean Modelling for Beginners*,
DOI 10.1007/978-3-642-00820-7_3, © Springer-Verlag Berlin Heidelberg 2009

3.2.2 *Contours and Contour Interval*

If a scalar varies spatially, this implies that it exhibits certain direction-dependent gradients. Gradients in the spatial distribution of a scalar therefore constitutes a vector field. To avoid confusion, consider the following example: A mountain can be characterised in terms of elevation of the ground above sea level. A map of this mountain can be produced by connecting points of the same elevation. These lines are called contours and the interval chosen between adjacent contour lines is called contour interval (Fig. 3.1). If one walks along elevation contours, there will be no change in elevation and it is impossible to climb or descent the mountain. In order to do so, contours have to be crossed at a certain angle. The steepest ascent or descent will be the one perpendicular to contour lines. The associated gradient, called bottom inclination or bottom slope, depends on direction and it is therefore a vector. The mountain is steepest where the spacing between contour lines is the closest.

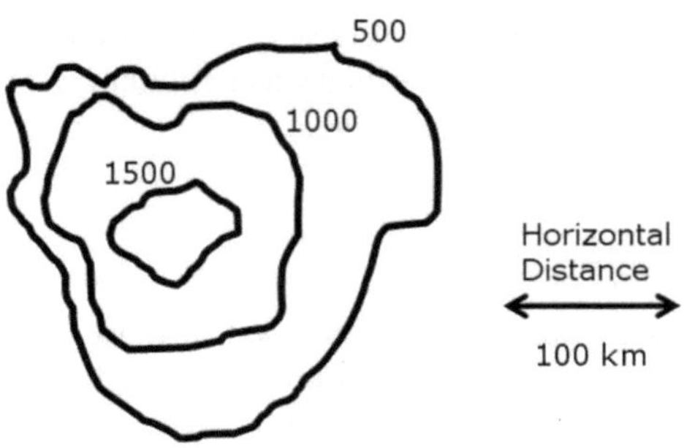

Fig. 3.1 Contours of elevation (m) above sea level of a mountain. The contour interval chosen is 500 m. Where are the steepest parts of the mountain? Can you climb the mountain by following a contour line?

3.3 Location and Velocity

3.3.1 *Location and Distance*

For horizontal distances of up to 100 km, or so, the spherical shape of the Earth's surface can be ignored and a plane rectangular coordinate system can be used. Fig. 3.2 shows such a so-called *Cartesian coordinate system*. The z-axis points upward and the coordinate surface at $z = 0$ m defines the (undisturbed) sea surface. With reference to the point-of-origin, defined by the coordinates $x = 0$, $y = 0$, and $z = 0$, any location can now by specified.

For convenience, specific locations are written in the form of (x_1, y_1, z_1). The location (10 m, 23 m, -10 m) is an example, where $z = -10$ m refers to a positive depth of 10 m below the undisturbed sea surface. The distance of a certain location from the point-of-origin is given by:

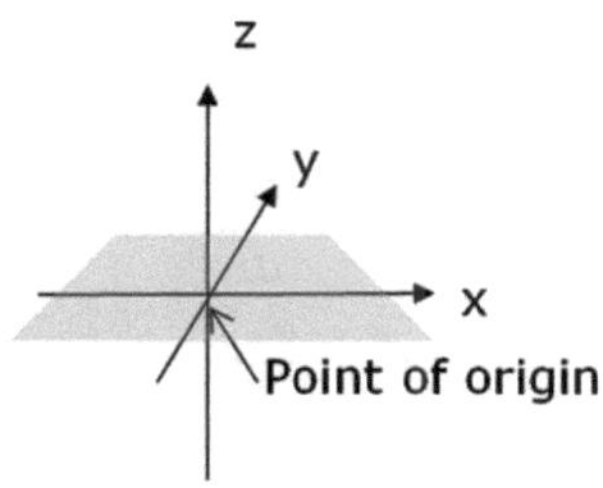

Fig. 3.2 The Cartesian coordinate system

$$D_1 = \sqrt{x_1^2 + y_1^2 + z_1^2}$$

The nice thing about vectors is that they can easily be appended by adding up the individual vector components. For instance, the sum of the location vectors (0, 5 m, −3 m) and (−2 m, −5 m, −2 m) gives a new location vector (−2 m, 0, −5 m). The distance between two locations (x_1, y_1, z_1) and (x_2, y_2, z_2) is given by:

$$D_{2-1} = \sqrt{(x_2 - x_1)^2 + (y_2 - y_1)^2 + (z_2 - z_1)^2}$$

In summary, location is specified by a vector that points from the point-of-origin to a certain spot in space. The magnitude of this vector is distance.

3.3.2 *Calculation of Distances with SciLab*

In the following example, we calculate the distance of the location (4.1 m, 2.6 m, −5.3 m) from the point-of-origin. Start SciLab and enter the following commands in the main control window:

```
x=4.1; y=2.6; z=−5.3; d1 = sqrt(x*x+y*y+z*z)
```

The square-root function is called "sqrt". Since no semicolon is used in the last line, the result of the calculation appears on the screen. The result (in metres) is:
d1 = 7.1874891

3.3.3 *Velocity*

Velocity is often expressed in terms of speed and direction of a solid or fluid parcel. Wind forecasts in local newspapers are an example of this. Instead of this, we will express velocity in component form in the x-, y- and z-directions. For convenience, we write this as (u, v, w). Since velocity is change of location of a moving parcel with time, we can also write:

$$u = \frac{dx^*}{dt}$$
$$v = \frac{dy^*}{dt} \quad (3.1)$$
$$w = \frac{dz^*}{dt}$$

where (x^*, y^*, z^*) is the location of a parcel. Speed is the magnitude of velocity and is given by:

$$\text{Speed} = \sqrt{u^2 + v^2 + w^2}$$

For example $(u, v, w) = (0.0, \; -10.0\,\text{m/s}, \; 0.0)$ refers to a parcel moving at a speed of 10 m/s into the negative y-direction; whereas (u,v,w) = (2 m/s, 2 m/s, 0.0) describes movement at a speed of $\sqrt{8} \approx 2.83$ m/s diagonally across the x-y plane.

3.4 Types of Motion

3.4.1 Steady-State Motions

A steady state is a situation in which currents do not show any time variations. This implies that there is a *balance* between all forces involved. Considerations of steady-state force balances are useful tools in geophysical fluid dynamics, leading to important relations such as the geostrophic balance, the thermal-wind relations or the Sverdrup balance, to be discussed below.

3.4.2 Waves

Waves are another type of motion being oscillations in time and space. Individual waves can be classified in terms of a *period* T (don not confuse this with temperature) and a *wavelength* λ (using the Greek symbol "lambda"). The wave period is the time lapse between successive peaks of a wave, whereas the wavelength is the distance between these peaks. One could measure the wave period with a stop watch, whereas the wavelength can be derived from instant photographs of the wave shape.

3.4.3 The Sinusoidal Waveform

It is convenient to use the sinusoidal function to describe waves in a mathematical manner. This function is based on radians and a complete cycle relates to a change of its argument by 2π (the Greek symbol "pi"), where π is about 3.1415. Accordingly, we can express a wave travelling in the x-direction as:

$$A(x,t) = A_o \sin\left[2\pi\left(\frac{x}{\lambda} - \frac{t}{T}\right)\right] \tag{3.2}$$

where A is a property experiencing oscillations such as sea level, A_o is the constant wave amplitude, being half the difference between maximum and minimum values of A, λ is wavelength, and T is wave period. This wave displays sinusoidal variations both in time and space. Equation (3.2) describes a wave void of variation in the y direction. Accordingly, wavefronts (crests and troughs) are parallel to each other. A wave like this is called a *plane wave*.

3.5 Visualisation of a Wave Using SciLab

3.5.1 A Simple Wave Made of Vertically Moving Bars

In order to visualise a wave, we consider a number of vertical bars sitting in a row next to each other. Bars rise or sink in a systematic pattern. The first bar gives the rhythm by moving up and down in a sinusoidal fashion. Any neighboring bar does the same but slightly delayed in time. Whenever a bar is above a certain horizon, it turns blue. When it is below this horizon, it turns red.

Figure 3.3 shows a snapshot of this wave. There is no lateral motion of any of the bars, but the pattern seems to move toward the right. The speed at which the pattern moves is called *phase speed* of a wave.

3.5.2 Sample Script

The SciLab script for this wave demonstration, called "WaveSim.sce", can be found in the folder "Miscellaneous" on the CD-ROM of this book. Before using this script, however, the reader should read the following brief introduction to SciLab scripting.

Fig. 3.3 Snapshot of organised wave motions by bars

3.5.3 The First SciLab Script

The following SciLab script displays the traditional message “Hello World” on the display.

```
// My first SciLab script
printf("Hello world");
```

The text editor within SciLab can be used to write this script, but make sure that file has the extension “.sce”. Once opened in SciLab, the script is executed by clicking on “Execute” of the top menu of the editor window. Alternatively, SciLab scripts can be run by entering “exec(‘c:\Folder\filename.sce’)” in the SciLab command window with reference to name and location of your script.

3.5.4 A Quick-Start to SciLab

SciLab programming commands are similar to those in FORTRAN, but there are some differences that are outlined in the following. In contrast to FORTRAN, SciLab programming is case sensitive. This implies that internal functions have to be called in lower case. FORTRAN “PROGRAM” and “END PROGRAM” statements are not needed in SciLab, and “ENDDO” and “ENDIF” statements are just called “end” statements in SciLab. Repetitive loops are called “for” loops in SciLab and an example is:

```
for i = 1:30; a(i) = i; end; // a simple "for" loop
```

Text after two forward slashes (//) are treated as comments. Values are allocated to arrays in two different ways. The following statement, for instance, defines a one-dimensional array with 30 elements each of a value of 10.0:

```
b(1:30) = 10.0;
```

Another option is the statement:

```
c = (0:1:30);
```

which produces a row vector of 31 elements with values c(0) = 0, c(1) = 1, …, c(30) = 30. A column vector can be produced by adding an apostrophe; that is,

```
c = (0:1:30)';
```

Without use of a semicolon, the result will be displayed in the SciLab command window. If statements in SciLab are written as:

```
if g > 0; f = sqrt(g); end;
```

SciLab can be used to write short simulation codes. Larger model codes, however, would take too long to run in SciLab and the preference in these cases should be FORTRAN. Specific SciLab plotting commands are explained in due course of this book.

3.5.5 *The First GIF Animation*

To produce a GIF animation watchable in most Web browsers, each picture frame produced in the SciLab script needs to be converted to an individual numbered GIF image. To this end, the script needs two additions. The following line needs to be added directly after the start of the animation loop:

```
scf(0); //set ID for graphic window
```

This allocates an identification number (0 here) to the graphics window. Within the animation loop, content of the graphics window can then be converted to a GIF file with:

```
if n < 10 then // n is the loop counter
    xs2gif(0,'MyFirst100'+string(n)+'.gif')
else
  if n < 100 then
    xs2gif(0,'MyFirst10'+string(n)+'.gif')
  else
    xs2gif(0,'MyFirst1'+string(n)+'.gif')
  end
end
```

The "string()" command converts a number into a string. Several string variables can be combined via simple addition. The "xs2gif(id,'name')" command saves the content of the graphics window with identification number "id" to a GIF file called 'name'. Instead of using running numbers of 1, 2, 3, 4, . . ., which can lead to some problems in the order of GIF files, we rather choose a sequence 1001, 1002, 1003, 1004, . . .

3.5.6 *Modified Animation Script*

The SciLab script including conversion of frames to GIF files, called "WaveSimGif.sce" can be found in the folder "Miscellaneous/Waves" on the CD-ROM accompanying this book.

3.5.7 Creation of an Animated GIF File

The "convert" program of ImageMagick can be used to convert the sequence of GIF files into a single animated GIF file. For example, a sequence of GIF files named img1001.gif, img1002.gif, img1003.gif, etc. can be composed in an animated GIF file (named anim.gif) by entering:

convert -delay 10 img.gif anim.gif*

in the Command Prompt window. The "-delay" option specifies the time lapse between frames in milliseconds. After moving the resultant animated GIF file to another safe location, individual GIF frame files should be immediately deleted to free up space on the computer.

3.5.8 Phase Speed

The *phase speed* of a wave is given by the ratio between wavelength and wave period:

$$c = \frac{\lambda}{T} \tag{3.3}$$

With this phase speed, Eq. (3.2) can be rewritten as:

$$A(x,t) = A_o \sin\left[\frac{2\pi}{\lambda}(x - ct)\right]. \tag{3.4}$$

A positive c makes the wave propagating into the positive x-direction, whereas a negative c makes it running into the opposite direction.

3.5.9 Dispersion Relation

The length and period of waves in fluids are in general not independent of each other. Instead of this, there is a relation between wave period and wavelength called the *dispersion relation*. The form of this relation varies with the situation. Waves of a wavelength long compared with the fluid depth behave much differently from waves of a wavelength short compared with the fluid depth. Section 4.2.5 presents the dispersion relation for long surface-gravity waves.

3.5.10 Superposition of Waves

The superposition of two or more waves of different period and/or wavelength can lead to various interference patterns such as a standing wave, being a wave of virtually zero phase speed. Interfering wave patterns travel with a certain speed, called *group speed* that can be different to the phase speeds of the contributing individual waves. Interference of storm-generated waves in the ocean can result in waves of gigantic wave heights (wave height is twice the wave amplitude) of >20 m, known as *freak waves*.

3.6 Exercise 2: Wave Interference

3.6.1 Aim

The aim of this exercise is to explore interferences that result from the superposition of two linear waves. To this end, SciLab will be used to calculate possible interferences pattern and to produce animations thereof.

3.6.2 Task Description

Consider the interference of two waves of the same amplitude of $A_o = 1$ m. The resultant wave can be described by:

$$A(x,t) = A_o \left\{ \sin\left[2\pi\left(\frac{x}{\lambda_1} - \frac{t}{T_1}\right)\right] + \sin\left[2\pi\left(\frac{x}{\lambda_2} - \frac{t}{T_2}\right)\right]\right\}$$

Using SciLab, the reader is asked to produce animations considering the following interference scenarios. In all scenarios, wave 1 has a period of $T_1 = 60$ s and a wavelength of $\lambda_1 = 100$ m. Choose period and wavelength of wave 2 from the following list:

Scenario 1: wavelength = 100 m; wave period = 50 s
Scenario 2: wavelength = 90 m; wave period = 60 s
Scenario 3: wavelength = 90 m; wave period = 50 s
Scenario 4: wavelength = 100 m; wave period = −60 s
Scenario 5: wavelength = 50 m; wave period = −30 s
Scenario 6: wavelength = 95 m; wave period = −30 s.

These scenarios describe a variety of interference patterns. Scenario 4, for instance, leads to a standing wave, being the result of two identical waves travelling in opposite directions. This is achieved by prescribing a negative value of the wave period for wave 2. Is this surprising how many different interference patterns can be created by superposition of just two waves. The reader is encouraged to experiment with other scenarios!

3.6.3 Sample Script

The SciLab script for this exercise, called "interference.sce" can be found in the folder "Exercise 2" on the CD-ROM.

3.6.4 A Glimpse of Results

Figure 3.4 shows a snapshot result for the last scenario. It took me a while to achieve what I wanted in terms of graphical display and, perhaps, the reader will come up with a better solution. It can be seen that, at some locations, wave disturbances add up to create a signal of doubled amplitude, whereas, in other regions, the waves compensate each other such that the resultant wave signal almost vanishes.

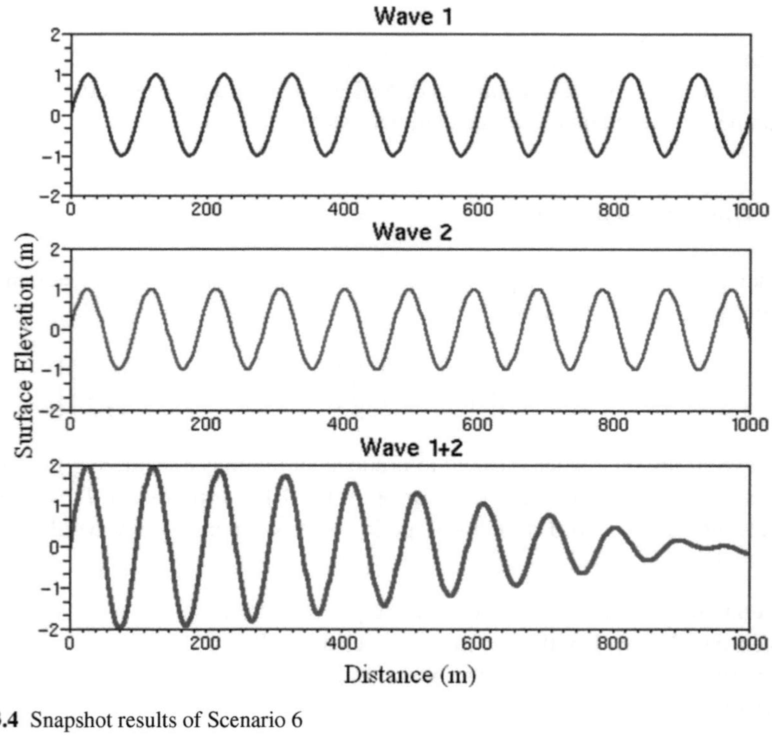

Fig. 3.4 Snapshot results of Scenario 6

3.6.5 A Rule of Thumb

The wave period should be resolved by at least 10 time steps. Otherwise details of the wave evolution can get lost. In the following SciLab script, I used 20 time steps in a period. Adequate choices of time steps and grid spacings play an important role in the modelling of dynamical processes in fluids.

3.7 Forces

3.7.1 What Forces Do

A non-zero force operates to change the speed and/or the direction of motion of a fluid parcel. In geophysical fluid dynamics, forces are conventionally expressed as forces per unit mass, being directly proportional to acceleration or deceleration of a fluid parcel. So, whenever the term "force" appears in the following, this actually should mean "force per unit mass".

In component form, a temporal change of the velocity vector can be formally written as:

$$\frac{du}{dt} = F^x$$
$$\frac{dv}{dt} = F^y$$
$$\frac{dw}{dt} = F^z$$

where (F^x, F^y, F^z) are the vector components of a force of certain magnitude and direction. For example, (0, 0, $-9.81\,\mathrm{m/s^2}$) is a force operating into the negative z-direction (downward). With two forces involved, the latter equations can be written as:

$$\frac{du}{dt} = F_1^x + F_2^x$$
$$\frac{dv}{dt} = F_1^y + F_2^y$$
$$\frac{dw}{dt} = F_1^z + F_2^z$$

Acceleration or deceleration results if any component of the resultant net force is different from zero. In the general case, the sum symbol "$\sum$" can be used to write:

$$\frac{du}{dt} = \sum_{i=1}^{m} F_i^x$$
$$\frac{dv}{dt} = \sum_{i=1}^{m} F_i^y, \qquad (3.5)$$
$$\frac{dw}{dt} = \sum_{i=1}^{m} F_i^z$$

where m is the number of forces involved in a process, and the index i points to a certain force.

3.7.2 Newton's Laws of Motion

Equations (3.5) already state the first two of *Newton's laws of motion* (Newton, 1687). Newton's first law of motion states that, in an absolute coordinate system void of any rotation or translation, both speed and direction of motion of a body remain unchanged in the absence of forces. If there is a force, on the other hand, there will be a certain change in motion. This is known as Newton's second law of motion.

3.7.3 Apparent Forces

Apparent forces come into play in a rotating coordinate system such as our Earth. One of these apparent forces is the *Coriolis force* that gives rise to circular weather patterns in the atmosphere, eddies in the ocean, or Jupiter's Red Spot.

3.7.4 Lagrangian Trajectories

Imagine that you sit on a fluid parcel of a certain temperature moving with the flow. The path along which you move is called a *Lagrangian trajectory*, based on work by Lagrange (1788). Without any heat exchange with the ambient fluid, the temperature of your fluid parcel remains constant and this feature can be formulated as:

$$\frac{dT}{dt} = 0 \tag{3.6}$$

where "T" is temperature and the "d" symbol now refers to a change of temperature *along the pathway of motion.*

3.7.5 Eulerian Frame of Reference and Advection

Instead of moving with the flow, you could stand still at a fixed location and measure changes in temperatures as the fluid moves past. This perspective is called the *Eulerian system*, based on work by Euler (1736). In this case, you would notice a change in temperature if a flow exists that carries differences (gradients) in temperature towards you. This process is called *advection.* In Cartesian coordinates, the effect of temperature advection can be expressed as:

$$\frac{\partial T}{\partial t} = -u\frac{\partial T}{\partial x} - v\frac{\partial T}{\partial y} - w\frac{\partial T}{\partial z} \tag{3.7}$$

This advection equation constitutes a *partial differential equation.*

3.7.6 Interpretation of the Advection Equation

The existence of both a flow and temperature gradients are essential ingredients in the advection process. The left-hand side of (3.7) is the temporal change in temperature measured at a fixed location. The appearance of minus signs on the right-hand side of (3.7) is not that difficult to understand. For simplicity, consider a flow running parallel to the x-direction. Recall that, per definition, u is positive if this flow component runs into the positive x-direction. Warming over time ($\partial T/\partial t > 0$) occurs with an increase of T in the x-direction in conjunction with a negative u. Warming also occurs with a positive u but a decrease of T in the x-direction. I am sure that the reader can work out scenarios leading to a local cooling.

In the absence of either flow or temperature gradients, Eq. (3.7) turns into:

$$\frac{\partial T}{\partial t} = 0 \tag{3.8}$$

This equation simply means that temperature does not show changes at a certain location. The important difference with respect to (3.6) is that this relation holds for a fixed location, whereas the other one was for an observer moving with the flow. Most ocean models use the Eulerian frame of reference.

3.7.7 The Nonlinear Terms

Flow can advect different properties such as gradients in temperature, salinity and nutrients, but also momentum; that is, the components of velocity itself. The resultant terms are called the *nonlinear terms*. These terms are included in Newton's second law of motion, if we express this in an Eulerian frame of reference, yielding:

$$\begin{aligned}
\frac{\partial u}{\partial t} + u\frac{\partial u}{\partial x} + v\frac{\partial u}{\partial y} + w\frac{\partial u}{\partial z} &= \sum_{i=1}^{m} F_i^x \\
\frac{\partial v}{\partial t} + u\frac{\partial v}{\partial x} + v\frac{\partial v}{\partial y} + w\frac{\partial v}{\partial z} &= \sum_{i=1}^{m} F_i^y \\
\frac{\partial w}{\partial t} + u\frac{\partial w}{\partial x} + v\frac{\partial w}{\partial y} + w\frac{\partial w}{\partial z} &= \sum_{i=1}^{m} F_i^z
\end{aligned} \tag{3.9}$$

The nonlinear terms are traditionally written on the left-hand side of the momentum conservation equations for they are no true forces.

3.7.8 Impacts of the Nonlinear Terms

The nonlinear terms are important in the dynamics of many processes. For instance, these terms are the reason for the existence of turbulence which makes mixing a soup with a spoon much more efficient than just waiting until the soup has mixed itself. The reader can also blame these terms for the unreliability of weather forecasts for longer than 5 days ahead.

3.8 Fundamental Conservation Principles

3.8.1 A List of Principles

There are several conservation principles that need to be considered when studying fluid motions. These are:

1. Conservation of momentum (Newton's laws of motion)
2. Conservation of mass
3. Conservation of interal energy (heat)
4. Conservation of salt.

In addition to this comes the so-called *equation of state* that links the field variables such as temperature and salinity to the density of the fluid. All these equations are coupled with each other, which makes the equations describing fluid motions a *coupled system of partial differential equations*.

3.8.2 Conservation of Momentum

Conservation of momentum is an expression of changes in flow speed and/or direction as a result of forces. The frictional stress imposed by winds along the sea surface acts as a boundary source term in the momentum equations. Friction at the sea flow acts as a sink term in these equations. Forces of relevance to fluids are explored in the next sections.

3.8.3 Conservation of Volume – The Continuity Equation

Water is largely incompressible, so that the mass of a given water volume cannot change much under compression. Conservation of mass can therefore be expressed in terms of conservation of volume. To understand this important concept, consider a virtual volume element (Fig. 3.5). For simplicity, we orientate this element in such a way that its face normals are parallel to the directions of the Cartesian coordinate system. The side-lengths of this box are δx, δy and δz, and the volume is $\delta V = \delta x \cdot \delta y \cdot \delta z$.

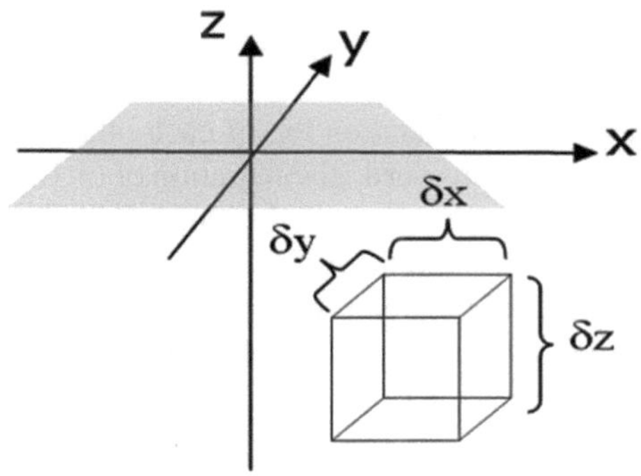

Fig. 3.5 A virtual control volume in a Cartesian coordinate system

Flow can enter or escape through any face of this volume element. Incompressibility of a fluid implies that all these individual inflows and outflows have to be balanced. Volume inflow or outflow is the product of the area of a face of our volume element and the flow component normal to it. The eastern and western faces span an area of $\delta y \cdot \delta z$ each and the relative volume change is given by $\delta u \cdot \delta y \cdot \delta z$, where δu is the difference of flow speed between both faces. This relative volume change can be reformulated as:

$$\delta u(\delta y \delta z) = \frac{\delta u}{\delta x}\delta x \delta y \delta z = \frac{\delta u}{\delta x}\delta V$$

Adding the contributions of the three pairs of opposite faces of the volume element and requesting this sum to be zero yields:

$$0 = \left(\frac{\delta u}{\delta x} + \frac{\delta v}{\delta y} + \frac{\delta w}{\delta z}\right)\delta V$$

Since δV is a positive and non-zero quantity, the final equation reads:

$$\frac{\delta u}{\delta x} + \frac{\delta v}{\delta y} + \frac{\delta w}{\delta z} = 0$$

The equation is valid for any finite volume and, accordingly, for a vanishingly small volume, which can be expressed by the partial differential equation

$$\frac{\partial u}{\partial x} + \frac{\partial v}{\partial y} + \frac{\partial w}{\partial z} = 0 \qquad (3.10)$$

being called the *continuity equation*. This equation constitutes the local form of volume conservation. One shortcoming when assuming an incompressible fluid is that acoustic waves in the fluid can no longer be described.

3.8.4 Vertically Integrated Form of the Continuity Equation

Small control volumes can be stocked on top of each other so that they extend the entire fluid column (Fig. 3.6). This vertical integration of (3.10) leads to a prognostic equation for the freely moving surface of the fluid, typically symbolized by the Greek letter η (spoken "eta"). The fluid surface will move up (or down) if there is a convergence (or divergence) of the depth-integrated horizontal flow.

In the absence of external sources or sinks of volume, such as precipitation (rainfall) or evaporation at the sea surface, the prognostic equation for η reads:

$$\frac{\partial \eta}{\partial t} = -\frac{\partial (h \langle u \rangle)}{\partial x} - \frac{\partial (h \langle v \rangle)}{\partial y} \tag{3.11}$$

where h is total fluid depth, and $\langle u \rangle$ and $\langle v \rangle$ are depth-averaged components of horizontal velocity. This equation is the *vertically integrated form* of the continuity equation for an incompressible fluid. The products $h \langle u \rangle$ and $h \langle v \rangle$ are depth-integrated lateral volume transports per unit width of the flow. Hence, the fluid level will change if the fluid column experiences a net lateral inflow or outflow of volume.

3.8.5 Divergence or Convergence?

There are two contributions that, if unbalanced, can change the surface level of the fluid. The first is associated with lateral variation of horizontal velocity, the other comes from flow in interaction with a sloping seafloor. These contributions can be quantified by applying the product rule for differentiation to the right-hand side terms of (3.11). In the x-direction, for instance, this rule gives:

$$\frac{\partial (h \langle u \rangle)}{\partial x} = h \frac{\partial \langle u \rangle}{\partial x} + \langle u \rangle \frac{\partial h}{\partial x}$$

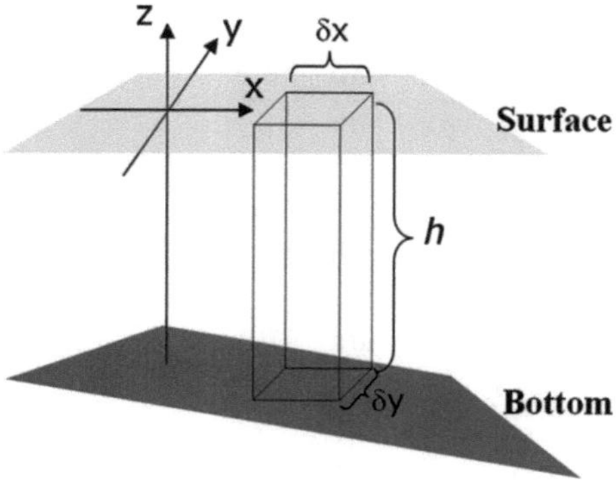

Fig. 3.6 A control volume in a Cartesian coordinate system extending from the bottom to the free surface of the fluid. Total fluid depth is h

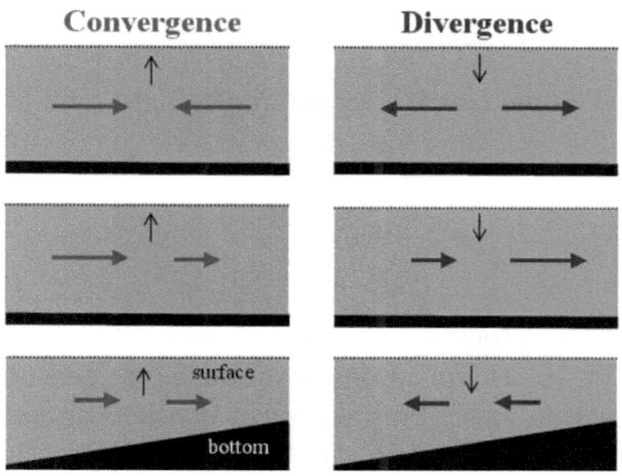

Fig. 3.7 Sketches of different flow fields leading to either lateral convergence or divergence of depth-averaged horizontal flow. *Vertical arrows* indicate the instant response of the sea surface

Each term is associated with either convergence or divergence of depth-averaged flow (Fig. 3.7), but it is the net effect of both terms that triggers the surface level to change.

3.8.6 The Continuity Equation for Streamflows

The continuity equation can also be applied to river flows or streamflows. Under the assumption of steady-state conditions, integration of the continuity equation over a cross-sectional area A of a river gives:

$$\langle u \rangle \cdot A = \langle u \rangle \cdot \langle h \rangle \cdot W = \text{constant}$$

where $\langle u \rangle$ is average flow speed, $\langle h \rangle$ is average depth, and W is width. Knowledge of the flow speed in a single transect of a river together with knowledge of river depth and width give the distribution of u along the full length of a river! The flow speed will increase in narrower river sections, if this is not compensated by a deepening of the river. Figure 3.8 illustrates this principle.

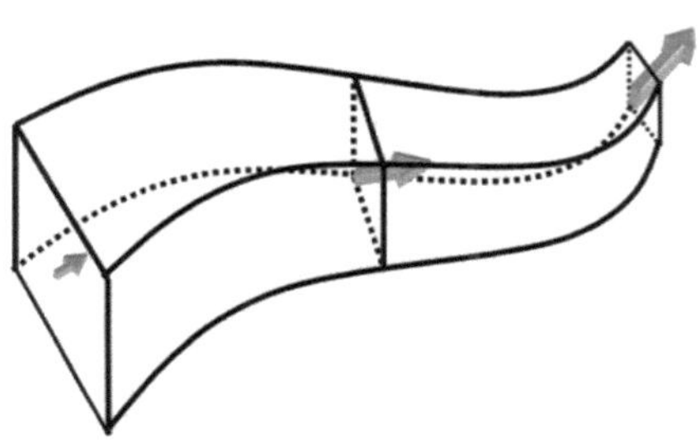

Fig. 3.8 Sketch of volume conservation in river flows. The flow speed increases in sections where the cross-sectional area of the river decreases

3.8.7 Density

Density, symbolised by the Greek letter ρ ("rho"), is the ratio of mass M over volume V:

$$\rho = \frac{M}{V} \tag{3.12}$$

Density has units of kilograms per cubic metres (kg/m^3). Density of seawater is in a range of 1025–1028 kg/m^3. Density variations are generally small ($< 0.5\%$) compared with the mean value. Freshwater has densities around 1000 kg/m^3. The density of seawater can be estimated by collecting a bucket of fluid and measure its weight over volume ratio.

3.8.8 The Equation of State for Seawater

Density of seawater depends on temperature, salinity and pressure. Pressure effects can be eliminated by converting the in-situ (on-site) temperature to that a water parcel would have when being moved adiabatically (without heat exchange with ambient fluid) to a certain reference pressure level.

Salinity is the mass concentration of dissolved salts in the water column. Seawater has typical salinities of 34–35 g/kg. Note that, in modern oceanography, salinity does not carry a unit, for it is defined and measured in terms of a dimensionless conductivity ratio. Numerical values of this *practical salinity*, however, are close to salt mass concentrations in g/kg.

To first-order approximation, it is often sufficient to calculate seawater density from a simple linear equation of state:

$$\rho(T, S) = \rho_o \left[1 - \alpha(T - T_o) + \beta(S - S_o)\right] \tag{3.13}$$

where mean density ρ_o refers to seawater density at reference temperature T_o and reference salinity S_o. The parameter α ("alpha") is the *thermal expansion coefficient* that attains a value of $2.5\times10^{-4}\,^\circ\mathrm{C}^{-1}$ at a temperature of 20°C. The salinity coefficient β ("beta") has values of 8×10^{-4} (no units). Oceanographers frequently use a quantity called σ_t ("sigma-t"). This is just the true density minus 1000 kg/m^3.

3.9 Gravity and the Buoyancy Force

3.9.1 Archimedes' Principle

Gravity is the gravitational pull toward the centre of Earth that a body would feel in the absence of a surrounding medium; that is, in a vacuum. Gravity acts downward in the Cartesian coordinate system and has vector components $(0, 0, -g)$, where $g = 9.81\ \text{m/s}^2$ is called *acceleration due to gravity*.

If an object of a certain mass M_{obj} and volume V_{obj} is released in the water column, instead of the full gravity force, it will experience only a reduced gravity force, called *buoyancy force*, that makes it sink at a slower rate or even rise in dependence of the density of the ambient medium. Imagine an air-filled plastic ball that will pop up if released in water.

According to *Archimedes' Principle*, the resultant buoyancy force is proportional to the difference between the object's mass with the mass of the fluid replaced by the object's volume. See Stein (1999) for a biography of Archimedes. The constant of proportionality is acceleration due to gravity. Accordingly, the resultant buoyancy force (per unit mass of the object) is given by:

$$\text{Buoyancy force} = -g\frac{(M_{obj} - M_{amb})}{M_{obj}}$$

where M_{amb} is the mass of the ambient fluid replaced by the object. After a few manipulations, this formula can be expressed in terms of densities, yielding:

$$\text{Buoyancy force} = -g\frac{(\rho_{obj} - \rho_{amb})}{\rho_{obj}} \tag{3.14}$$

where ρ_{amb} is the density of the ambient fluid. Note that the buoyancy force has a negative sign and is directed downward if the object's density exceeds that of the surrounding fluid. Now it will be not that difficult for the reader to tell why ships made from steel usually stay at the sea surface (unless capsizing).

3.9.2 Reduced Gravity

In physical oceanography, the negative of the buoyancy force (per unit mass) is commonly termed *reduced gravity*. This quantity carries the symbol g' and is given by:

$$g' = g\frac{(\rho_{obj} - \rho_{amb})}{\rho_{obj}} \tag{3.15}$$

It should be noted that buoyancy is not only restricted to objects of a solid skin. The object of interest can be a fluid parcel itself.

3.9.3 Stability Frequency

Vertical density gradients can be expressed in terms of the *stability frequency* N, traditionally called Brunt – Väisälä frequency with appreciation of early works by

Brunt (1927) and Väisälä (1925), that is defined by:

$$N^2 = -\frac{g}{\rho_o}\frac{\partial \rho}{\partial z} \tag{3.16}$$

where ρ_o is mean density. Linear density stratification can therefore be expressed as:

$$\rho(z) = \rho_o \left(1 + \frac{N^2}{g}|z|\right) \tag{3.17}$$

where ρ_o is surface density.

3.9.4 Stable, Neutral and Unstable Conditions

The situation of $N^2 > 0$ refers to a *stably stratified* fluid column. *Neutral conditions* are given for $N^2 = 0$. The fluid column is *statically unstable* (dense fluid above light fluid) with $N^2 < 0$. This situation cannot exist for long as it triggers convection operating to stir the water column. Convection is involved in the motion of bubbles arising when heating a pot of water or soup from below.

3.10 Exercise 3: Oscillations of a Buoyant Object

3.10.1 Aim

The aim of this exercise is to predict the pathway of a buoyant parcel in a stratified water column using FORTRAN for the prediction and SciLab for visualisation of the result.

3.10.2 Task Description

For the following exercise, we assume that density in a 100-m deep water column increases linearly with depth. The surface density is 1025 kg/m^3 and the stability frequency squared is taken as $N^2 = 10^{-4}$ s^{-2}, so that density increases to 1026 kg/m^3 at the bottom. The task for the reader is now to predict the motion path of an object of 1025.5 kg/m^3 in density when released at a depth of, say, 80 m.

3.10.3 Momentum Equations

For simplicity, we assume that there is only motion in the vertical. Accordingly, the momentum equations (3.9) reduce to a single equation:

$$\frac{dw_{\text{obj}}}{dt} = -g\,\frac{(\rho_{\text{obj}} - \rho_{\text{amb}})}{\rho_{\text{obj}}} \tag{3.18}$$

The buoyancy force on the right-hand side of the latter equation varies in magnitude and sign in dependence of the object's location. On the other hand, the location of the object, z_{obj}, changes owing to its vertical speed w_{obj} according to:

$$\frac{dz_{\text{obj}}}{dt} = w_{\text{obj}} \tag{3.19}$$

The latter two equations are coupled with each other, for the object's location determines the density found in the ambient fluid and, hence, the magnitude of the buoyancy force.

3.10.4 Code Structure

The code has three parts:

1. a predictor for vertical speed w_{obj}
2. a predictor for the new location z_{obj}
3. a calculator of the ambient oceanic density with respect to the object's location.

3.10.5 Finite-Difference Equations

In finite-difference form, Eq. (3.18) can be written as:

$$w_{\text{obj}}^{n+1} = w_{\text{obj}}^{n} - \Delta t\; g\;(\rho_{\text{obj}} - \rho_{\text{amb}})/\rho_{\text{obj}} \tag{3.20}$$

where n is the time level and Δt is the time step chosen. Equation (3.19) can be written as:

$$z_{\text{obj}}^{n+1} = z_{\text{obj}}^{n} + \Delta t \cdot w_{\text{obj}}^{n+1} \tag{3.21}$$

The use of time level $(n+1)$ for vertical speed w_{obj} in the latter equation just means that input into this equation comes from the predicted value of w_{obj} from the previous equation. In other words, step (3.20) has to come before step (3.21). The time step is set to $\Delta t = 1$ s.

3.10.6 Initial and Boundary Conditions

The initial location of the object set to $z_{\text{obj}}^0 = -80\,\text{m}$. The initial vertical speed w_{obj}^0 is set to zero. *Boundary conditions* need to be specified in addition to this to avoid that the objects digs into the seafloor or pops out of the water column. The easiest solution is to make these boundaries impermeable.

3.10.7 Sample Code and Animation Script

The FORTRAN 95 source code for this exercise, called "buoyant.f95", and the SciLab animation script, called "buoyant.sce" can be found in the folder "Exercise 3" on the CD-ROM. The file "info.txt" contains additional information.

3.10.8 Discussion of Results

Figure 3.9 displays the numerical solution to Exercise 3. Initially, the object is lighter compared with the ambient fluid and experiences a positive (upward) buoyancy force. Hence, the object becomes subject to upward acceleration and its vertical speed increases until the object reaches its equilibrium level at a depth of 50 m. This takes about 2.5 h.

It overshoots this level, for it takes some time before deceleration has reduced the object's speed again to zero. When this occurs, however, the object finds itself at a depth horizon of around 20 m in a less dense environment. It experiences a negative buoyancy force and, accordingly, is subject to downward acceleration. Again, it moves past its equilibrium density level. This overshooting is a form of *inertia* and the resultant movement of the object is a vertical oscillation about its equilibrium density level. The period of this oscillation is slightly above 10 min.

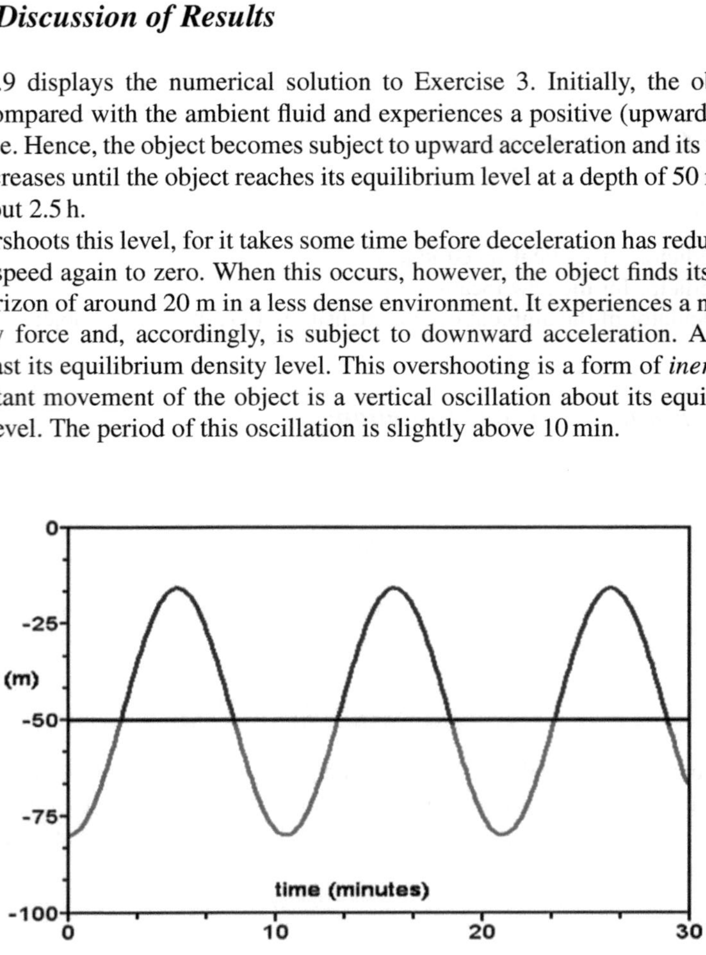

Fig. 3.9 Location of the buoyant object as a function of time. The 50-m depth corresponds to the equilibrium density horizon of the object

3.10.9 Analytical Solution

Equations (3.18) and (3.19) can be combined to yield a single equation with the argument z_{obj}. This equation reads:

$$\frac{dw_{\text{obj}}}{dt} = \frac{d^2 z_{\text{obj}}}{dt^2} = -g\frac{(\rho_{\text{obj}} - \rho_{\text{amb}})}{\rho_{\text{obj}}} \tag{3.22}$$

For convenience, we define z^* as the distance of the object from its equilibrium density level. The buoyancy force is proportional to the distance from this equilibrium level, and, with aid of (3.17), we can write the latter equation as:

$$\frac{d^2 z^*}{dt^2} = -N^2 \, z^* \tag{3.23}$$

Accordingly, we seek a function whose second temporal derivative gives the same function times a constant N^2 and a sign reversal. Only one particular type of functions does this – the sinusoidal function – and the solution is:

$$z^*(t) = z_o^* \, \cos(Nt) \tag{3.24}$$

where z_o^* is the initial distance from the equilibrium level. The period of this wave is related to the stability frequency as $T = 2\pi/N$. We yield $T = 628.3\,\text{s} = 10.47\,\text{mins}$ for settings of Exercise 3. The prediction is close to this analytical result. The reader is encouraged to test the solution (3.24) for other choices of stability frequency N, object densities or initial displacement distances. Vertical speed evolves according to:

$$w_{\text{obj}} = \frac{dz^*}{dt} = z_o^* \, N \, \sin(Nt) = w_o \, \sin(Nt)$$

where $w_o = z_o^* \, N$ is the maximum speed that the object attains as it crosses its density equilibrium level. In our example, $z_o^* = 30$ m and $N = 0.01\ \text{s}^{-1}$ give a maximum vertical speed of around 30 cm/s.

3.10.10 Inclusion of Friction

Under the assumption that the object is subject to friction in proportion to its speed, Eq. (3.18) can be expanded as:

$$\frac{dw_{\text{obj}}}{dt} = -g \, \frac{(\rho_{\text{obj}} - \rho_{\text{amb}})}{\rho_{\text{obj}}} - Rw_{\text{obj}} \tag{3.25}$$

where the friction coefficient R has units of 1/s. Using an implicit approach for this term, the finite difference form of this equation reads:

$$\frac{w_{\mathrm{obj}}^{n+1} - w_{\mathrm{obj}}^{n}}{\Delta t} = -g\,\frac{\rho_{\mathrm{obj}} - \rho_{\mathrm{amb}}}{\rho_{\mathrm{obj}}} - R w_{\mathrm{obj}}^{n+1} \tag{3.26}$$

which can be reorganised to yield:

$$w_{\mathrm{obj}}^{n+1} = \left[w_{\mathrm{obj}}^{n} - \Delta t \cdot g\,(\rho_{\mathrm{obj}} - \rho_{\mathrm{amb}})/\rho_{\mathrm{obj}} \right] /(1 + R\Delta t) \tag{3.27}$$

Figure 3.10 shows the solution for $R = 0.002\ \mathrm{s}^{-1}$ describing a *damped oscillation*. Only a few changes are required in the FORTRAN code to achieve this. First, the friction parameter needs to be added in the declaration section:

```
REAL :: R ! friction parameter
```

Second, this parameter is specified with:

```
R = 0.002 ! value of friction parameter
```

Last, the friction force is added in the momentum equation with:

```
WOBJN=(WOBJ+dt*BF)/(1.+R*dt) ! predict new vertical speed
```

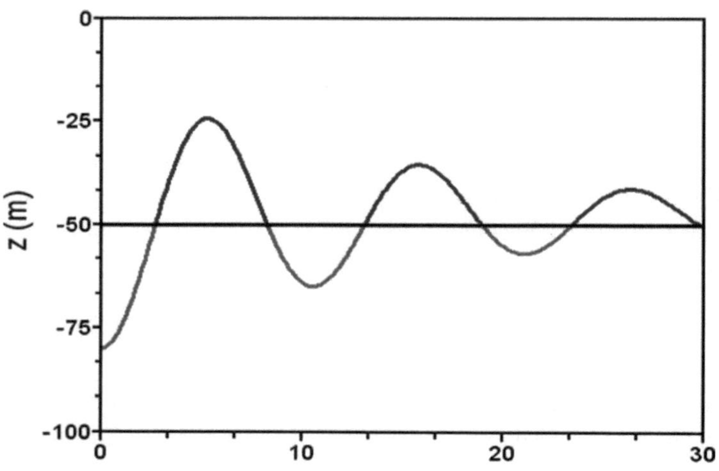

Fig. 3.10 Same as Fig. 3.9, but with inclusion of friction

3.10.11 *Additional Exercises for the Reader*

Repeat this exercise with the inclusion of friction and vary the friction coefficient in a range between 0.001 s^{-1} and 0.5 s^{-1}. Advanced readers can also try to implement a quadratic friction term of the form $-R \mid W \mid W$, where R has now units of 1/m. Use a semi-implicit approach for this term; that is $-R \mid W^n \mid W^{n+1}$, and choose different values of R. Use the FORTRAN function "ABS()" to calculate the absolute value $\mid W^n \mid$.

3.11 The Pressure-Gradient Force

3.11.1 *The Hydrostatic Balance*

For a fluid at rest, the downward acting gravity force is balanced by an upward acting pressure-gradient force. This balance, called the *hydrostatic balance* or *hydrostatic approximation*, can be written as:

$$0 = -\frac{1}{\rho}\frac{\partial P}{\partial z} - g \tag{3.28}$$

where P is pressure, z is vertical coordinate, ρ is local density, and g is acceleration due to gravity. The minus sign in the pressure-gradient terms arises to make this term positive if pressure decreases with height.

3.11.2 *Which Processes are Hydrostatic?*

It can be shown that processes of a horizontal scale large compared with their vertical scale are hydrostatic. Otherwise the dynamics are said to be *nonhydrostatic*, which implies that the pressure field is modified by the flow. This book exclusively deals with hydrostatic processes.

3.11.3 *The Hydrostatic Pressure Field in the Ocean*

Hydrostatic pressure in the ocean has three contributions: atmospheric pressure, pressure excess or deficit owing to elevated or lowered sea level, and pressure owing to the density stratification in the ocean itself. Atmospheric pressure has no impact, for the sea surface tends to adjust instantaneously to atmospheric pressure variations, such that the pressure below the sea surface remains virtually the same. This is known as the *inverted barometer effect*. The pressure field associated with the mean density and a plane sea level has no dynamical consequences for it is void of horizontal gradients.

3.11.4 Dynamic Pressure in the Ocean

Dynamic pressure is the pressure that remains if we exclude pressure contributions void of dynamical consequences. The vertical distribution of dynamic (hydrostatic) pressure in the ocean is given by:

$$P(z^*) = \rho_o\, g\, \eta + g \int_{z^*}^{\eta} \rho' dz \tag{3.29}$$

where z^* is vertical location in the water column, η is sea level elevation, and ρ' is density anomaly compared with mean density ρ_o. The integral in the latter equation is proportional to the weight anomaly of the water column above location z^*. With exclusion of inactive pressure contributions, the hydrostatic relation takes the form:

$$0 = -\frac{\partial P}{\partial z} - \rho' g$$

It is rather reduced gravity than gravity that makes up dynamically relevant pressure-gradients in the fluid's interior.

3.11.5 The Horizontal Pressure-Gradient Force

One of the dominant forces producing fluid motion is the horizontal pressure-gradient force. This force arises from a slanting surface of the fluid and/or horizontal gradients in density. The pressure-gradient force (per unit mass) has horizontal vector components of:

$$-\frac{1}{\rho}\frac{\partial P}{\partial x} \quad \text{and} \quad -\frac{1}{\rho}\frac{\partial P}{\partial y} \tag{3.30}$$

Minus signs are required because the pressure-gradient force acts from high to low pressure.

3.11.6 The Boussinesq Approximation

Density variations in the ocean are <1% compared with mean density. Hence, the horizontal pressure-gradient force can be approximated by:

$$-\frac{1}{\rho_o}\frac{\partial P}{\partial x} \quad \text{and} \quad -\frac{1}{\rho_o}\frac{\partial P}{\partial y}$$

where ρ_o is a constant mean density, and the hydrostatic balance can be rewritten as:

$$0 = -\frac{1}{\rho_o}\frac{\partial P}{\partial z} - g'$$

where reduced gravity is given by $g' = \rho'/\rho_o g$. This simplification is known as the *Boussinesq approximation* in appreciation of early work by *Boussinesq (1903)*.

3.11.7 The Case of Uniform Density

In a fluid of uniform density, the horizontal pressure-gradient force turns into:

$$-g\frac{\partial \eta}{\partial x} \quad \text{and} \quad -g\frac{\partial \eta}{\partial y}$$

which is yielded when inserting (3.29) with $\rho' = 0$ into (3.30). In the absence of density variations, this force is the same throughout the fluid column. We can therefore expect that the resultant flow is also depth-independent, a feature referred to as *barotropic flow*. In contrast to this, pressure gradients associated with lateral density difference in the ocean interior triggers horizontal flow that changes with depth. Such a flow is called *baroclinic flow*. Figure 3.11 shows sketches of barotropic and baroclinic flows.

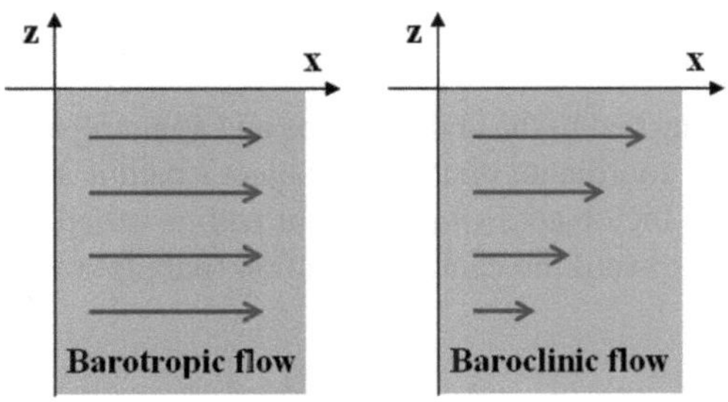

Fig. 3.11 Examples of barotropic and baroclinic flows

3.12 The Coriolis Force

3.12.1 Apparent Forces

Newton's laws of motion are valid in a fixed coordinate system, that is one that doesn't rotate or translate. These laws imply that in the absence of forces, objects follow a straight path with unchanged speed. In a rotating coordinate system, however, straight paths appear as curved paths (Fig. 3.12). If we want to apply Newton's laws of motion in rotating coordinates, this implies the existence of apparent forces.

There are two different apparent forces involved in the observed curved path of the object, namely the *centrifugal force* and the *Coriolis force (Coriolis, 1835)*. In order to understand the Coriolis force, we need to understand the centrifugal force first.

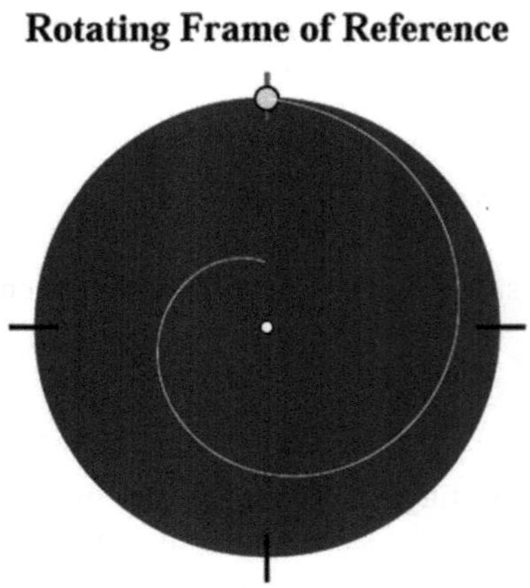

Fig. 3.12 Appearance of straight path (*white line*) of an object (*ball*) for an observer on a rotating turntable. The inner end of the *white line* shows the starting position of the object. The SciLab script "Straight Path.sce" in the folder "Miscellaneous/Coriolis Force" of the CD-ROM produces an animation

3.12.2 The Centripetal Force and the Centrifugal Force

Consider an object attached to the end of a rope and spinning around with a rotating turntable. In the fixed coordinate system, the object's path is a circle (Fig. 3.13) and the force that deflects the object from a straight path is called the *centripetal force*. This force is directed toward the centre of rotation (parallel to the rope) and, hence,

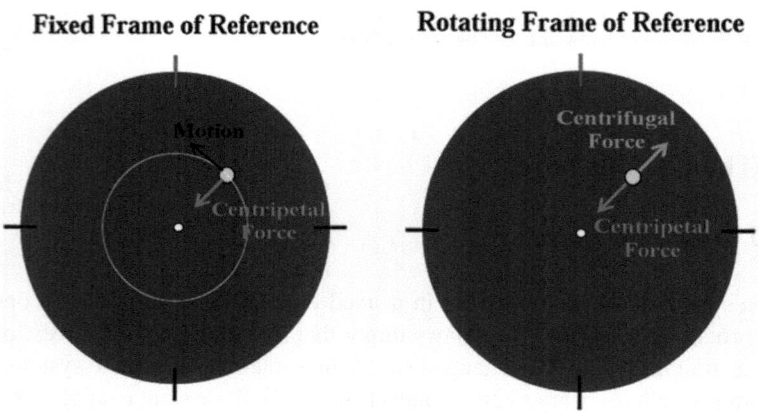

Fig. 3.13 *Left panel* In the fixed frame of reference, an object attached to a rope and rotating at the same rate as the turntable experiences only the centripetal force. *Right panel*: The object appears stationary in the rotating frame of reference, which implies a balance between the centripetal force and the centrifugal force. The SciLab script "Centripetal Force.sce" in the folder "Miscellaneous/Coriolis Force" of the CD-ROM produces an animation

operates perpendicular to the object's direction of motion. The centripetal force (per unit mass), which is a true force, is given by:

$$\text{Centripetal force} = -\Omega^2\, r \tag{3.31}$$

where r is the object's distance from the centre of the turntable, and $\Omega = 2\pi/T$ is the rotation rate with T being the rotation period. Per definition, the rotation rate Ω is positive for anticlockwise rotation and negative for clockwise rotation. When releasing the object, it will fly away on a straight path with reference to the fixed frame of reference.

In the rotating frame of reference, on the other hand, the object remains at the same location and is therefore not moving at all. Consequently, the centripetal force must be balanced by another force of the same magnitude but acting in the opposite direction. This apparent force – the *centrifugal force* – is directed away from the centre of rotation. Accordingly, the centrifugal force is given by:

$$\text{Centrifugal force} = +\Omega^2\, r \tag{3.32}$$

When releasing the object, an observer in the rotating frame of reference will see the object flying away on a curved path – similar to that shown in Fig. 3.12.

3.12.3 Derivation of the Centripetal Force

The speed of any object attached to the turntable is the distance travelled over a time span. Paths are circles with a circumference of $2\pi r$, where r is the distance from the centre of rotation, and the time span to complete this circle is the rotation period. Accordingly, the speed of motion is given by:

$$v = \frac{2\pi}{T} r = \Omega r. \tag{3.33}$$

During rotation, the speed of parcels remains the same, but the direction of motion and thus the velocity changes (Fig. 3.14). The similar triangles in Fig. 3.14 give the relation $\delta v/v = \delta L/r$. Since δL is given by speed multiplied by time span, this relation can be rearranged to yield the centripetal force (per unit mass):

$$\frac{dv}{dt} = -\frac{v^2}{r}, \tag{3.34}$$

where the minus sign has been included since this force points toward the centre of rotation. Equation (3.31) follows, if we finally insert (3.33) into the latter equation.

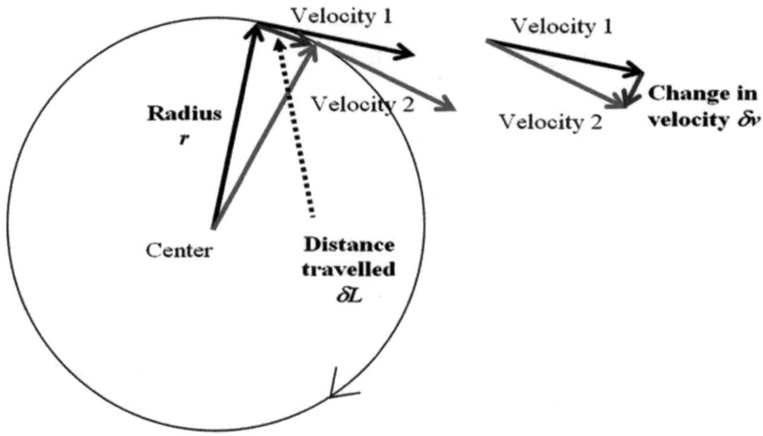

Fig. 3.14 Changes of location and velocity of a parcel on a turntable

3.12.4 The Centrifugal Force in a Rotating Fluid

Consider a circular tank filled with fluid on a rotating turntable. Letting the tank rotate at a constant rate for a long time, all fluid will eventually rotate at the same rate as the tank. In this steady-state situation, the fluid surface attains a noneven shape, as sketched in Fig. 3.15. The final shape of the fluid surface is determined by a balance between the centrifugal force and a centripetal force, that, in our rotating fluid, is provided by a horizontal pressure-gradient force provided by a slanting fluid surface. This balance of forces reads:

$$-g\frac{\partial \eta}{\partial r} = -\Omega^2 r \tag{3.35}$$

where r is the radial distance from the centre of the tank.

The analytical solution of the latter equation is:

$$\eta(r) = \frac{1}{2}\frac{\Omega^2}{g}r^2 - \eta_o, \tag{3.36}$$

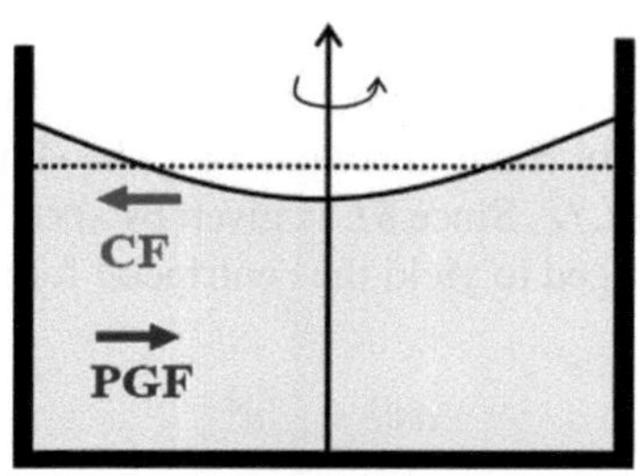

Fig. 3.15 Sketch of the steady-state force balance between centrifugal force (CF) and pressure-gradient force (PGF) in a rotating fluid. The *dashed line* shows the equilibrium surface level for the nonrotating case

where the constant η_o can be determined from the requirement that the total volume of fluid contained in the tank has to be conserved (if the tank is void of leaks). The tank's rotation leads to a parabolic shape of the fluid surface and it is essentially gravity (via the hydrostatic balance) that operates to balance out the centrifugal force. The latter balance is valid for all fluid parcels in the tank. The pathways of fluid parcels are circles in the fixed frame of reference. The observer in the rotating system, however, will not spot any movement at all.

3.12.5 Motion in a Rotating Fluid as Seen in the Fixed Frame of Reference

With reference to a fixed frame of reference, fluid parcels in the rotating tank exclusively feel the centripetal force provided by the pressure-gradient force. In the absence of relative motion, fluid parcels describe circular paths. How does the trajectory of a fluid parcel look like, if we give it initially a push of a certain speed into a certain direction? The momentum equations governing this problem are given by:

$$\frac{dU}{dt} = -\Omega^2 X \text{ and } \frac{dV}{dt} = -\Omega^2 Y \tag{3.37}$$

where (X, Y) refers to a location and (U, V) to a velocity in the fixed coordinate system. On the other hand, the location of our fluid parcel simply changes according to:

$$\frac{dX}{dt} = U \text{ and } \frac{dY}{dt} = V \tag{3.38}$$

Owing to rotation, velocities in the fixed and rotating reference systems are not the same. Instead of this, it can be shown that they are related according to:

$$U = u - \Omega\, y \text{ and } V = v + \Omega\, x \tag{3.39}$$

where (x, y) refers to a location and (u, v) to a velocity in the rotating coordinate system.

3.12.6 Parcel Trajectory

Before reviewing the analytical solution, we employ a numerical code (see below) to predict the pathway of a fluid parcel in a rotating fluid tank as appearing in the fixed frame of reference. To this end, we consider a fluid tank, 20 km in diameter, rotating at a rate of $\Omega = -0.727 \times 10^{-5}\ \text{s}^{-1}$, which corresponds to clockwise rotation with a period of 24 h.

At location $X = x = 0$ and $Y = y = 5$ km, a disturbance is introduced such that the fluid parcel obtains a relative speed of $u_o = 0.5$ m/s and $v_o = 0.5$ m/s. In the fixed coordinate frame, the initial velocity is $U_o = 0.864$ m/s and $V_o = 0.5$ m/s.

The results show that the resultant path of the fluid parcel is elliptical (Fig. 3.16). With a closer inspection of selected snapshots of the animation (Fig. 3.17), we can also see that the fluid parcel comes closest to the rim of the tank twice during one full revolution of the fluid tank. This finding, which is simply the result of the elliptical path, is the important clue to understand why so-called *inertial oscillations*, described below, have periods half that associated with the rotating coordinate system.

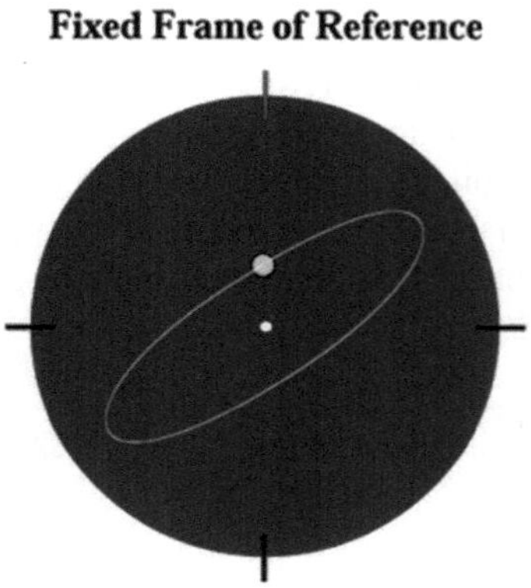

Fig. 3.16 Trajectory of motion (*white line*) for one complete revolution of a clockwise rotating fluid tank as seen in the fixed frame of reference. The SciLab script "Traject" in the folder "Miscellaneous/Coriolis Force" of the CD-ROM produces an animation

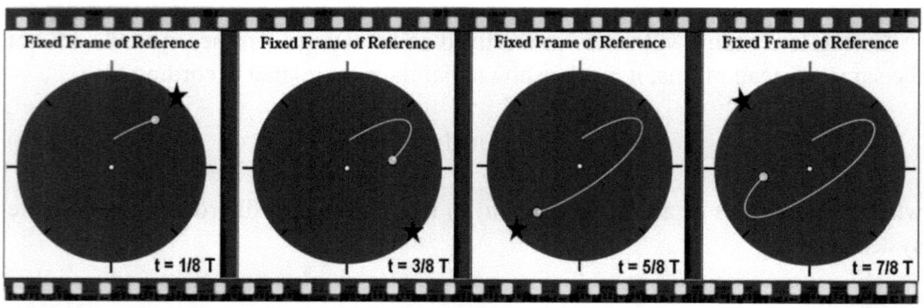

Fig. 3.17 Same as Fig. 3.16, but shown for different time instances of the simulation. The tank rotates in a clockwise sense. The *star* denotes a fixed location at the rim of the rotation tank

3.12.7 Numerical Code

In finite-difference form, the momentum equations (3.37) can be written as:

$$U^{n+1} = U^n - \Delta t \cdot \Omega^2 X^n \quad \text{and} \quad V^{n+1} = V^n - \Delta t \cdot \Omega^2 Y^n \tag{3.40}$$

where n is time level and Δt is time step. The trajectory of our fluid parcel can be predicted with:

$$X^{n+1} = X^n + \Delta t \cdot U^{n+1} \text{ and } Y^{n+1} = Y^n + \Delta t \cdot V^{n+1} \tag{3.41}$$

Again, predictions from the momentum equations are inserted into the latter equations as to yield an update of the locations. I decided to tackle this problem entirely with SciLab without writing a FORTRAN simulation code.

3.12.8 Analytical Solution

Equations (3.37) and (3.38) can be combined to yield:

$$\frac{d^2X}{dt^2} = -\Omega^2 X \text{ and } \frac{d^2Y}{dt^2} = -\Omega^2 Y \tag{3.42}$$

The solution of these equations that satisfies initial conditions in terms of location and velocity are given by:

$$X(t) = X_o \cos(\Omega t) + \frac{U_o}{\Omega} \sin(\Omega t) \tag{3.43}$$

$$Y(t) = Y_o \cos(\Omega t) + \frac{V_o}{\Omega} \sin(\Omega t) \tag{3.44}$$

This solution describes the trajectory of a parcel along an elliptical path. In the absence of an initial disturbance ($u = 0$ and $v = 0$), and using (3.39), the latter equations turn into:

$$X(t) = X_o \cos(\Omega t) - Y_o \sin(\Omega t)$$
$$Y(t) = Y_o \cos(\Omega t) + X_o \sin(\Omega t)$$

which is the trajectory along a circle of radius $\sqrt{X_o^2 + Y_o^2}$, as expected.

3.12.9 The Coriolis Force

We can now reveal the Coriolis force by translating the trajectory seen in the fixed frame of reference (see Fig. 3.16), described by (3.43) and (3.44), into coordinates of the rotating frame of reference. The corresponding transformation reads:

$$x = X \cos(\Omega t) + Y \sin(\Omega t) \tag{3.45}$$

$$y = Y \cos(\Omega t) - X \sin(\Omega t). \tag{3.46}$$

Figure 3.18 shows the resultant flow path as seen by an observer in the rotating frame of reference. Interestingly, the fluid parcel follows a circular path and completes the

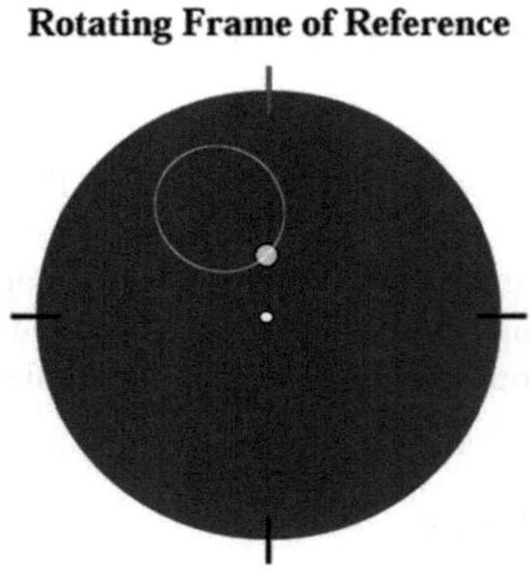

Fig. 3.18 Pathway of an object that experiences the Coriolis force in a clockwise rotating fluid. The SciLab script "Coriolis Force Revealed.sce" in the folder "Miscellaneous/Coriolis Force" of the CD-ROM produces an animation

circle twice while the tank revolves only once about its centre. Accordingly, the period of this so-called *inertial oscillation* is 0.5 T, known as *inertial period*, with T being the rotation period of the fluid tank.

Rather than working in a fixed coordinate system, it is more convenient to formulate the Coriolis force from the viewpoint of an observer in the rotating frame of reference. In the absence of other forces, it can be shown that inertial oscillations are governed by the momentum equations:

$$\frac{\partial u}{\partial t} = +2\Omega\ v \ \text{ and } \ \frac{\partial v}{\partial t} = -2\Omega\ u \qquad (3.47)$$

The Coriolis force acts perpendicular to the direction of motion and the factor of 2 reflects the fact that inertial oscillations have a period half that of the rotating frame of reference. If a parcel is pushed with an initial speed of u_o into a certain direction, it can also be shown that its resultant path is a circle of radius $u_o/(2\,|\Omega)|)$. With an initial speed of about 0.7 m/s and $|\Omega| = 0.727\times10^{-5}\ \text{s}^{-1}$, as in the above example, this *inertial radius* is about 4.8 km.

3.13 The Coriolis Force on Earth

3.13.1 The Local Vertical

In rotating fluids at rest, the centrifugal force is compensated by pressure-gradient forces associated with slight modification of the shape of the fluid surface. On the rotating Earth, this leads to a minor variation of the gravity force by less than 0.4%. The *local vertical* at any geographical location is now defined as the coordinate axis aligned at right angle to the equilibrium sea surface. This implies that, for a

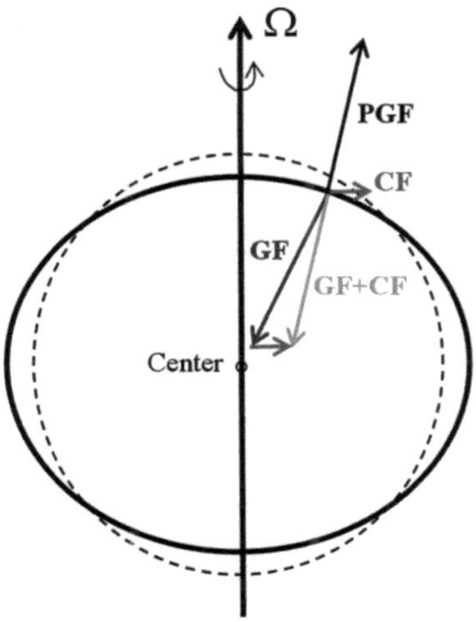

Fig. 3.19 Balances of forces on a rotating Earth fully covered with seawater in a state at rest. The gravity force (GF) is directed toward the Earth's centre. The centrifugal force (CF) acts perpendicular to the rotation axis. The pressure-gradient force (PGF) balances the combined effects of GF and CF. The local vertical is parallel to PGF

state at rest, the pressure-gradient force along this local vertical perfectly balances the combined effect of the gravity force and the centrifugal force (Fig. 3.19).

3.13.2 The Coriolis Parameter

Owing to a discrepancy between the orientations of the rotation axis of Earth and the local vertical, the magnitude of the Coriolis force becomes dependent on geographical latitude and Eqs. (3.47) turn into:

$$\frac{\partial u}{\partial t} = +fv \text{ and } \frac{\partial v}{\partial t} = -fu \tag{3.48}$$

where $f = 2\Omega\sin(\varphi)$, with φ being geographical latitude, is called the *Coriolis parameter*. The Coriolis parameter changes sign between the northern and southern hemisphere and vanishes at the equator. This variation of the Coriolis parameter can be explained by a modification of the centripetal force in dependence of the orientation of the local vertical (Fig. 3.20). Consequently, the period of inertial oscillations is $T = 2\pi/|f|$ and it depends exclusively on geographical latitude. It is 12 hours at the poles and goes to infinity near the equator. The radius of inertial circles is given by $u_o/|f|$. Inertial oscillations attain a clockwise sense of rotation in the northern hemisphere and describe counterclockwise paths in the southern hemisphere.

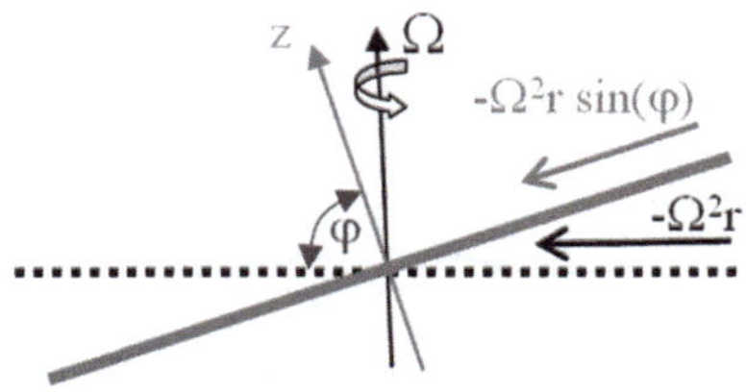

Fig. 3.20 The centripetal force for a variation of the orientation of local vertical

3.13.3 The f-Plane Approximation

The curvature of the Earth's surface can be ignored on spatial scales of 100 km, to first-order approximation. Hence, on this scale, we can place our Cartesian coordinate system somewhere at the sea surface with the z-axis pointing into the direction of the local vertical and use a constant Coriolis parameter (Fig. 3.21). The constant value of f is defined with respect to the point-of-origin of our coordinate system. This configuration is called the *f-plane approximation*.

3.13.4 The Beta-Plane Approximation

The curved nature of the sea surface can still be ignored on spatial scales of up to a 1000 km (spans about 10° in latitude), if the Coriolis parameter is described by a constant value plus a linear change according to:

$$f = f_o + \beta\ y \tag{3.49}$$

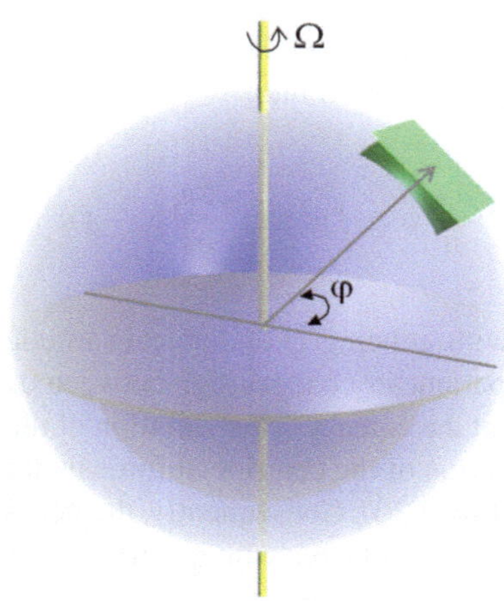

Fig. 3.21 The sketch gives an example of a f-plane. The Coriolis parameter is given by $f = 2\Omega \sin(\varphi)$, where φ is geographical latitude of the centre of the plane

In this approximation, β is the meridional variation of the Coriolis parameter with a value of $\beta = 2.2 \times 10^{-11}\mathrm{m}^{-1}\mathrm{s}^{-1}$ at mid-latitudes, and y is the distance in metres with respect to the centre of the Cartesian coordinates system defining f_o. Note that y becomes negative for locations south of this centre. Equation (3.49) is known as the *beta-plane approximation.*

A spherical coordinate system is required to study dynamical processes of length-scales greater than 1000 km. A discussion of such processes, however, is beyond the scope of this book.

3.14 Exercise 4: The Coriolis Force in Action

3.14.1 Aim

The aim of this exercise is to predict the pathway of a non-buoyant fluid parcel in a rotating fluid subject to the Coriolis force.

3.14.2 First Attempt

With the settings detailed in Sect. 3.12.6, we can now try to simulate the Coriolis force in a rotating fluid by formulating (3.48) in finite-difference form as:

$$u^{n+1} = u^n + \Delta t f v^n \quad \text{and} \quad v^{n+1} = v^n - \Delta t f u^n$$

Locations of our fluid parcel are predicted with:

$$x^{n+1} = x^n + \Delta t\ u^{n+1}, \quad \text{and} \quad y^{n+1} = y^n + \Delta t\ v^{n+1}$$

The result of this scheme is disappointing and, instead of the expected circular path, shows a spiralling trajectory (Fig. 3.22). Obviously, there is something wrong here. The problem here is that the velocity change vector is perpendicular to the actual velocity at any time instance, so that the parcel ends up outside the inertial circle (Fig. 3.23). This error grows with each time step of the simulation and the speed of the parcel increases gradually over time, which is in conflict with the analytical solution. This explicit numerical scheme is therefore *numerically unstable* and must not be used.

3.14.3 Improved Scheme 1: the Semi-Implicit Approach

Circular motion is achieved by formulating (3.48) in terms of a semi-implicit scheme:

$$u^{n+1} = u^n + 0.5\ \alpha(v^n + v^{n+1}) \text{ and } v^{n+1} = v^n - 0.5\ \alpha(u^n + u^{n+1}) \tag{3.50}$$

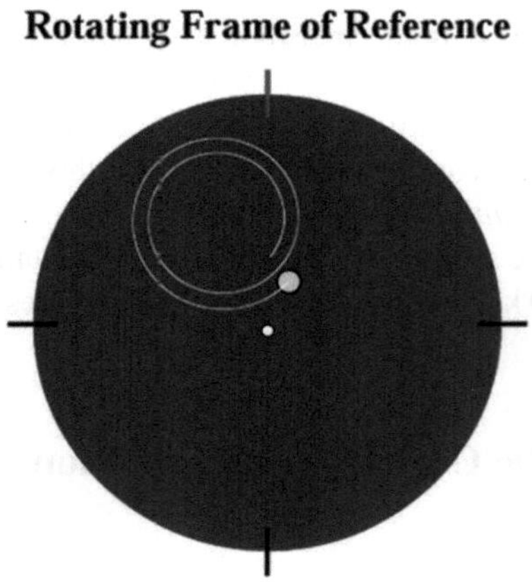

Fig. 3.22 First attempt to simulate the Coriolis force

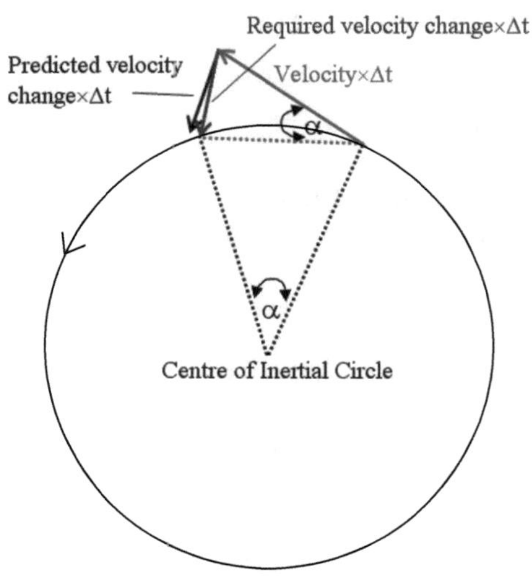

Fig. 3.23 Illustration of the error inherent with the explicit scheme

where $\alpha = \Delta t f$. A cross-combination of the latter equations gives:

$$u^{n+1} = \left[(1-\beta)u^n + \alpha v^n\right]/(1+\beta) \tag{3.51}$$

$$v^{n+1} = \left[(1-\beta)v^n - \alpha u^n\right]/(1+\beta) \tag{3.52}$$

where $\beta = 0.25\ \alpha^2$. This scheme requires numerical time steps small compared with the rotation period; that is, $|\alpha| << 1$, otherwise the period of the parcel's circular motion will differ from the true value. This semi-implicit scheme is widely used by modellers. It is worth noting that a fully-implicit scheme would lead to inward spiralling of trajectories and a gradual decrease in speed, which is certainly not intended.

3.14.4 Improved Scheme 2: The Local-Rotation Approach

The Coriolis force operates at a right angle to velocity and does not change the speed of motion, only the direction. Hence, this feature can be simulated by a local rotation of the velocity vector; that is,

$$u^{n+1} = \cos(\alpha)u^n + \sin(\alpha)v^n, \tag{3.53}$$

$$v^{n+1} = \cos(\alpha)v^n - \sin(\alpha)u^n \tag{3.54}$$

From geometric considerations, the rotation angle can be determined at $\alpha = 2 \arcsin(0.5\Delta t f)$. For $\Delta t\,|f| << 1$, this can be approximated by $\alpha \approx \Delta t f$.

3.14.5 Yes!

Figure 3.24 shows the results using the semi-implicit scheme demonstrating that we are now able to successfully simulate inertial oscillations in a rotating fluid. With this code, the reader is encouraged to explore inertial oscillations for a variety of situations, such as for different geographical locations and different initial locations and speeds. The code also includes a formulation of the local-rotation approach with $\alpha \approx \Delta t f$ that can be selected via the parameter "mode".

3.14.6 Sample Code and Animation Script

The FORTRAN code for this exercise, called "Coriolis.f95", and the SciLab script, "Coriolis.sce" can be found in the folder "Exercise 4" on the CD-ROM. The file "info.txt" contains additional information.

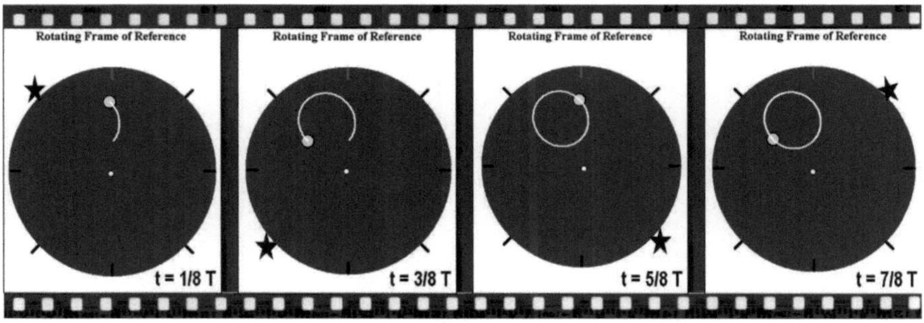

Fig. 3.24 Snapshots of the trajectory (*white line*) of a water parcel subject to the Coriolis force as predicted with the semi-implicit approach. The *star* denotes a reference location in the fixed frame of reference

3.14.7 Inertial Oscillations

The aim of this section is to predict the pathway of a non-buoyant parcel floating with an ambient uniform flow (U_o, V_o) subject to a series of abrupt wind events. The momentum equations governing this problem can be formally written as:

$$\frac{\partial u}{\partial t} = +fv + \frac{\partial u_f}{\partial t} \tag{3.55}$$

$$\frac{\partial v}{\partial t} = -fu + \frac{\partial v_f}{\partial t} \tag{3.56}$$

where u_f and v_f are forcing terms assumed to be uniform in space. The pathway of the fluid parcel can then be calculated from:

$$\frac{dx}{dt} = U_o + u \quad \text{and} \quad \frac{dy}{dt} = V_o + v \tag{3.57}$$

To illustrate this process, we consider an ambient flow with velocity components of (U_o, V_o) = (5 cm/s, 5 cm/s), corresponding a uniform northeastward flow. In addition to this, we consider three abrupt events in which the relative flow changes speed and direction. For a total simulation time of 6 days, the first event occurs at time zero and produces a change of the relative flow of (Δu_f, Δv_f) = (10 cm/s, 0 cm/s). The second event takes place at day 2 and produces a flow change of (Δu_f, Δv_f) =

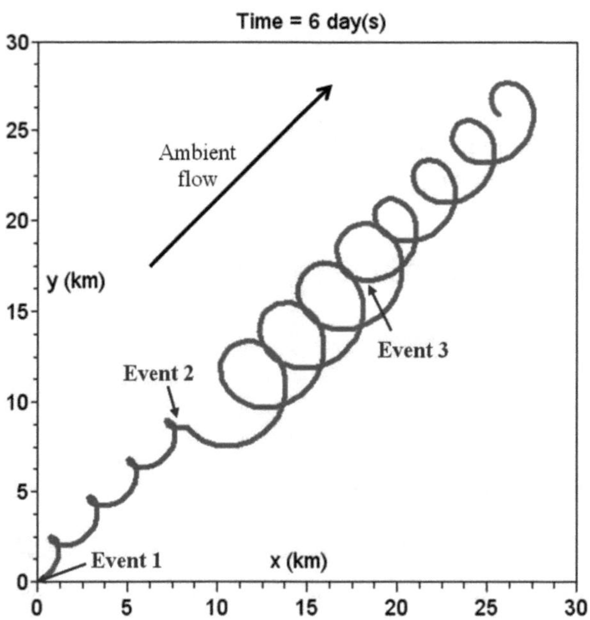

Fig. 3.25 Pathway of a fluid parcel carried by an ambient flow and subject to inertial oscillations

(10 cm/s, 0 cm/s). The last event happens at day 4 with $(\Delta u_f, \Delta v_f) = (0.0$ cm/s, 10.0 cm/s).

Figure 3.25 shows the resultant flow path of a water parcel starting from the point-of-origin. As you can see, the parcel is carried by the ambient flow into a northeastward direction. Superimposed are inertial oscillations. If you play around with the intensity and direction of abrupt flow disturbances, you will learn that these disturbances can either destroy or amplify pre-existing inertial oscillations.

3.14.8 Sample Code and Animation Script

The modified FORTRAN code, called "InerOsci.f95", and the SciLab script, "InerOsci.sce", can be found in the folder "Miscellaneous/Inertial Oscillations" on the CD-ROM.

3.15 Turbulence

3.15.1 Laminar and Turbulent Flow

Laminar flow is a smooth flow that doesn't exhibit a great deal of irregular motions. We can test whether a flow is laminar by throwing a stick into the water. If the stick does not swirl around much, the flow is said to be *laminar*. Conversely, if the stick shows irregular movements or even flips over, the flow is obviously *turbulent*.

3.15.2 The Reynolds Approach

Fluctuations of a physical property, such as temperature, are the trace of turbulence. Accordingly, we can express an observed quantity that we call ψ (the Greek symbol "psi") in terms of a mean value plus fluctuations (Fig. 3.26):

$$\psi = \langle \psi \rangle + \psi' \tag{3.58}$$

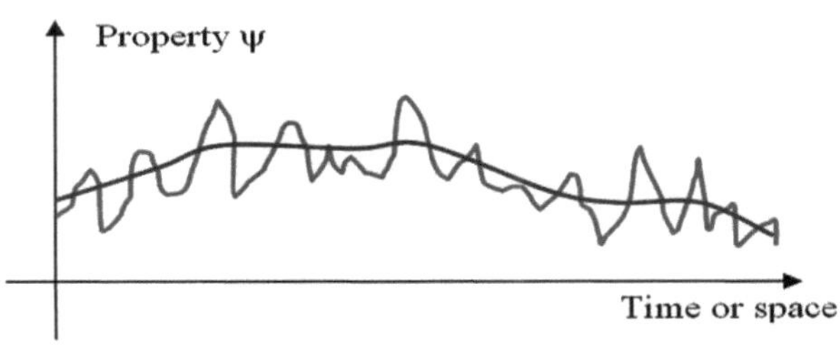

Fig. 3.26 Observed values of a physical property (*wiggly line*) can be expressed in terms of a mean value (*smooth line*), averaged over a certain time and/or space interval, plus fluctuations (difference between both lines)

where $\langle \psi \rangle$ is a mean value and ψ' are fluctuations. This approach, proposed by Osborne Reynolds (1895), is the key ingredient in the theory of turbulence.

3.15.3 What Causes Turbulence?

There are two principle causes of turbulence in a fluid. One source of turbulence are *shear flows*. Shear flow is a flow that displays speed variations perpendicular to its movement direction. Figure 3.27 gives examples of shear flows. Owing to frictional effects, strong shear flows establish in the vicinity of fluid boundaries. The other cause of turbulence is convective mixing that occurs in case of unstable vertical density stratification. A stable density stratification, on the other hand, can weaken or even suppress turbulence.

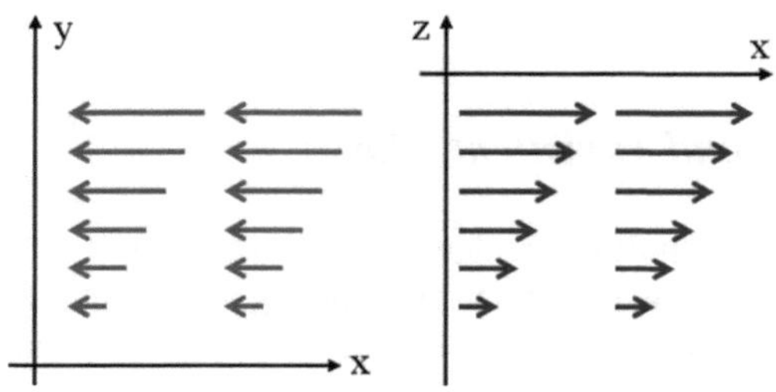

Fig. 3.27 Examples of shear flows

3.15.4 The Richardson Number

The transition between laminar and turbulent flow in a stratified, vertical shear flow can be characterised by means of a nondimensional number – the Richardson number. This number compares turbulent energy production/dissipation associated with vertical density stratification and turbulent energy production associated with vertical shear of a mean horizontal flow. For a mean flow $\langle u \rangle$ running into the x-direction, for example, this number is given by:

$$Ri = \frac{N^2}{(\partial \langle u \rangle / \partial z)^2} \tag{3.59}$$

where N^2 is the stability frequency of the ambient (mean) density field. Theory and experimental studies suggest that turbulence is created whenever Ri falls below a threshold value of around 1/4 = 0.25. Richardson (1920) was the first to derive this instability criterion.

3.15.5 Turbulence Closure and Turbulent Diffusion

A *turbulence closure* is a mathematical expression that relates fluctuations of a variable to properties of the mean flow. Some sort of turbulence closure is required in a finite-difference model, for we cannot resolve processes of periods shorter than the time step or lengthscales smaller than the grid spacing.

A conventional approach is the assumption that turbulence operates as a mixing agent to smooth sharp gradients in a property field and to describe the effect of this by means of a diffusion equation, written as:

$$\frac{\partial \psi}{\partial t} = \frac{\partial}{\partial x}\left(K_x \frac{\partial \psi}{\partial x}\right) + \frac{\partial}{\partial y}\left(K_y \frac{\partial \psi}{\partial y}\right) + \frac{\partial}{\partial z}\left(K_z \frac{\partial \psi}{\partial z}\right) \tag{3.60}$$

where ψ is a property of interest, such as temperature, and K_x, K_y, and K_z are certain coefficients parameterising the effect of turbulence. These coefficients can carry different values in the case of direction-dependent turbulence. For instance, you can stir a soup such that the cream added mixes horizontally rather than in the vertical. For scalar fields, such as temperature or salinity, the coefficients are called *eddy diffusivities*. In the case of momentum diffusion, they are called *eddy viscosities*.

3.15.6 Prandtl's Mixing Length

Vertical velocity shear is a source of turbulence in a fluid. In the absence of density stratification, vertical eddy viscosity A_z is in proportion to the magnitude of velocity shear via the relationship:

$$A_z = L^2 \left| \frac{\partial u}{\partial z} \right| \tag{3.61}$$

where the lengthscale L – *Prandtl's mixing length* – (Prandtl, 1925) is a measure of the diameter of turbulent elements, also called vortices. The latter equation, being a simplified turbulence closure, requires information on the size of vortices. More advanced turbulence closures, not detailed here, include effects of density stratification via some dependency on the Richardson number.

3.15.7 Interpretation of the Diffusion Equation

To understand the process of diffusion, we consider a depth-varying temperature field subject to turbulent diffusion in the vertical represented by a constant eddy diffusivity. Then, Eq. (3.60) can be written as:

$$\frac{\partial T}{\partial t} = K_z \frac{\partial^2 T}{\partial z^2} \tag{3.62}$$

This equation implies that the temperature will locally change as long as the temperature profile is curved. In other words, the temperature distribution approaches a steady state once the a uniform vertical gradient in temperature is established in the fluid column. The value of turbulent diffusivity K_z determines the time it takes for this steady state to establish. Obviously, the resultant vertical temperature gradient depends crucially on heat fluxes across boundaries. In the absence of such heat fluxes, the steady-state temperature field can only be uniform (well-mixed) throughout the fluid.

3.16 The Navier–Stokes Equations

3.16.1 Complete Set of Equations

The *Navier–Stokes equations* (Navier, 1822; Stokes, 1845) comprise a set of coupled conservation equations required to describe motions in fluids. These equations consist of the momentum equations, the continuity equation (expressing conservation of volume), advection-diffusion equations for field variables such as temperature and salinity, and the equation of state. The momentum equations can be expressed by:

$$\begin{aligned} \frac{\partial u}{\partial t} + \mathrm{Adv}(u) - fv &= -\frac{1}{\rho_o}\frac{\partial P}{\partial x} + \mathrm{Diff}(u) \\ \frac{\partial v}{\partial t} + \mathrm{Adv}(v) + fu &= -\frac{1}{\rho_o}\frac{\partial P}{\partial y} + \mathrm{Diff}(v) \\ \frac{\partial w}{\partial t} + \mathrm{Adv}(w) &= -\frac{1}{\rho_o}\frac{\partial P}{\partial z} - \frac{\rho'}{\rho_o}g + \mathrm{Diff}(w) \end{aligned} \tag{3.63}$$

where (u, v, w) is the velocity vector, t is time, (x, y, z) is the location vector in the Cartesian coordinate system, f is the Coriolis parameter, P is dynamic pressure, g' is reduced gravity, the operator Adv() denotes the nonlinear terms, and Diff() refers to the diffusion terms.

The continuity equation in its local form is given by:

$$\frac{\partial u}{\partial x} + \frac{\partial v}{\partial y} + \frac{\partial w}{\partial z} = 0 \tag{3.64}$$

For a linear equation of state and the assumption that eddy diffusivities are the same for temperature and salinity, we can formulate an advection-diffusion equation for density anomalies, called the *density conservation equation*, that can be formulated as:

$$\frac{\partial \rho'}{\partial t} + \mathrm{Adv}(\rho') = \mathrm{Diff}(\rho') \tag{3.65}$$

3.16.2 Boundary Conditions for Oceanic Applications

Wind stress operates as a frictional force at the sea surface. The associated boundary condition is given by:

$$\left(A_z \frac{\partial u}{\partial z}\right)_{z=0} = \frac{\tau_x^{\text{wind}}}{\rho_o} \text{ and } \left(A_z \frac{\partial v}{\partial z}\right)_{z=0} = \frac{\tau_y^{\text{wind}}}{\rho_o} \tag{3.66}$$

where ρ_o is surface density. The components of the wind-stress vector are given by:

$$\tau_x^{\text{wind}} = \rho_{\text{air}} C_d \; U\sqrt{U^2+V^2} \text{ and } \tau_y^{\text{wind}} = \rho_{\text{air}} C_d \; V\sqrt{U^2+V^2} \tag{3.67}$$

where ρ_{air} is air density, C_d is the nondimensional *wind-drag coefficient* with values of the order of $1.1\text{–}1.5\times10^{-3}$, and U and V are horizontal components of the wind vector measured at a height of 10 m above sea level.

Bottom friction is usually treated by either a linear or a quadratic approach. The linear approach reads:

$$\frac{\tau_x^{\text{bot}}}{\rho_o} = \left(A_z \frac{\partial u}{\partial z}\right)_{z=-H} = r_{\text{lin}}\; u \text{ and } \frac{\tau_y^{\text{bot}}}{\rho_o} = \left(A_z \frac{\partial v}{\partial z}\right)_{z=-H} = r_{\text{lin}}\; v \tag{3.68}$$

where H is total depth of the water column, the friction parameter r_{lin} has units of metres per second, and (u, v) is the lateral flow in vicinity of the seafloor. The quadratic bottom-friction law is given by:

$$\frac{\tau_x^{\text{bot}}}{\rho_o} = r\; u\sqrt{u^2+v^2} \text{ and } \frac{\tau_y^{\text{bot}}}{\rho_o} = r\; v\sqrt{u^2+v^2} \tag{3.69}$$

where r is a nondimensional *bottom-drag coefficient*. Other boundary conditions include the vertically integrated form of the continuity equation (3.11) that we need to predict sea-level elevation and associated barotropic pressure gradients. Source and sink boundary terms associated with volume changes such as precipitation need to be added to this equation, if required. Also required are boundary conditions describing density changes owing to surface density fluxes.

3.17 Scaling

3.17.1 The Idea

The idea behind the scaling theory is that we can estimate the relative significance of terms in the Navier–Stokes equations by using typical magnitudes or *scales* of flow variables. For instance, take a surface wave in the ocean that has a certain period T and wavelength λ. If we want to model this wave, an obvious question to ask is whether we need to include the nonlinear terms and the Coriolis force in the momentum equations, or can either or both of these terms be neglected under certain circumstances?

3.17.2 Example of Scaling

Consider a wave propagating into the x-direction of a flow speed u described by the momentum equation:

$$\frac{\partial u}{\partial t} + u\frac{\partial u}{\partial x} - fv = -\frac{1}{\rho_o}\frac{\partial P}{\partial x} \tag{3.70}$$

Notice that a number of terms have already been dropped from the full equation under the assumption they are small compared with the remaining terms. What is the relative magnitude of terms listed in this equation? Which term is large and which one is small? Scales will help here. For a wave, the characteristic scales of motion are flow speed (U_o), wave period T, and wavelength λ. Hence, the order of magnitude of the first term in the above equation can be estimated at:

$$\frac{\partial u}{\partial t} \propto \frac{U_o}{T} \tag{3.71}$$

The magnitude of the second term – the nonlinear term – can be estimated at:

$$u\frac{\partial u}{\partial x} \propto U_o\frac{U_o}{\lambda} = \frac{U_o^2}{\lambda} \tag{3.72}$$

The ratio between these estimates gives the *Froude number*, introduced by William Froude (1874):

$$\frac{U_o^2/\lambda}{U_o/T} = \frac{U_o T}{\lambda} = \frac{U_o}{c} \tag{3.73}$$

where c is the phase speed of the wave, which is different from the horizontal speed that a fluid parcel experiences. The conclusion is that nonlinear terms can be ignored in wave problems if the phase speed of the wave exceeds by far the lateral speed of fluid parcels.

With a similar approach, we can estimate the relative importance of the Coriolis force that has a scale of fU_o. A comparison between this scale with that of the temporal change (3.71) gives:

$$\frac{U_o/T}{f\,U_o} = \frac{1}{f\,T} = \frac{T_i}{T} \tag{3.74}$$

where T_i is the inertial period, which is the *temporal Rossby number*. Consequently, we can neglect the Coriolis force for waves of a period much shorter compared with the inertial period. If we now deal with a wave of $T << T_i$ and $c >> U_o$, the governing momentum equation reduces to:

$$\frac{\partial u}{\partial t} = -\frac{1}{\rho_o}\frac{\partial P}{\partial x} \tag{3.75}$$

It is obvious that scaling considerations can lead to substantail simplification of the dynamical equations governing a certain process. It is also fascinating that the nature of the Navier–Stokes equations varies in dependence on the scales of a dynamical process.

Chapter 4
Long Waves in a Channel

Abstract This chapter introduces the reader to the modelling of layered flows in one-dimensional channel applications including a simple flooding algorithm. Practical exercises address a variety of processes including shallow-water surface waves, tsunamis and interfacial waves in a multi-layer fluid.

4.1 More on Finite Differences

4.1.1 Taylor Series

The value of a function in vicinity of given location x can be expressed in form of a *Taylor series* (Taylor, 1715) as:

$$f(x+\Delta x) = f(x) + \Delta x \frac{\partial f}{\partial x} + \frac{\Delta x^2}{1 \cdot 2} \frac{\partial^2 f}{\partial x^2} + \frac{\Delta x^3}{1 \cdot 2 \cdot 3} \frac{\partial^3 f}{\partial x^3} + \cdots \tag{4.1}$$

The essence of this series is that the neighboring value can be reconstructed by means of the value at x plus a linear correction using the slope of f at location x plus a higher-order correction involving the curvature of f at x and so on. Accordingly, the first derivative of a function can be approximated by:

$$\frac{\partial f}{\partial x} \approx \frac{f(x+\Delta x) - f(x)}{\Delta x} \tag{4.2}$$

but we have to admit that this expression is not 100% accurate owing to neglection of higher-order terms. Alternatively, the Taylor series can be written as:

$$f(x-\Delta x) = f(x) - \Delta x \frac{\partial f}{\partial x} + \frac{\Delta x^2}{1 \cdot 2} \frac{\partial^2 f}{\partial x^2} - \frac{\Delta x^3}{1 \cdot 2 \cdot 3} \frac{\partial^3 f}{\partial x^3} + \cdots \tag{4.3}$$

J. Kämpf, *Ocean Modelling for Beginners*,
DOI 10.1007/978-3-642-00820-7_4,

On the basis of this, another approximation of the first derivative of a function is:

$$\frac{\partial f}{\partial x} \approx \frac{f(x) - f(x - \Delta x)}{\Delta x} \tag{4.4}$$

The third option is to take the sum of both Taylor series, yielding:

$$\frac{\partial f}{\partial x} \approx \frac{f(x + \Delta x) - f(x - \Delta x)}{2\Delta x} \tag{4.5}$$

There are three different options of expressing the first derivative of a function in terms of a finite difference.

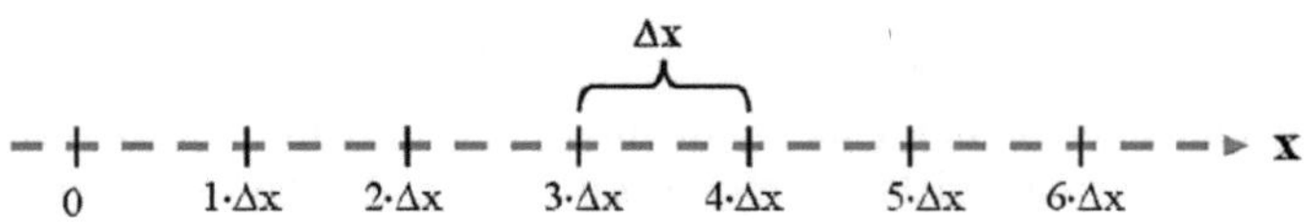

Fig. 4.1 Example of equidistant grid spacing. Distance is given by $k \cdot \Delta x$, where k is the cell index and Δx is grid spacing

4.1.2 Forward, Backward and Centred Differences

With the choice of equidistant grid spacing and index notation (Fig. 4.1), we can formulate the different finite-difference forms of the first derivative of a function as:

$$\frac{\partial f}{\partial x} \approx \frac{f_{k+1} - f_k}{\Delta x} \tag{4.6}$$

called *forward difference*, or

$$\frac{\partial f}{\partial x} \approx \frac{f_k - f_{k-1}}{\Delta x} \tag{4.7}$$

called *backward difference*, or

$$\frac{\partial f}{\partial x} \approx \frac{f_{k+1} - f_{k-1}}{2\Delta x} \tag{4.8}$$

called *centred difference*.

4.1.3 Scheme for the Second Derivative

The sum of the Taylor series (4.1) and (4.3) gives an approximation of the second derivative of a function:

$$\frac{\partial^2 f}{\partial x^2} \approx \frac{f(x+\Delta x) - 2f(x) + f(x-\Delta x)}{(\Delta x)^2} = \frac{f_{k+1} - 2f_k + f_{k-1}}{(\Delta x)^2} \tag{4.9}$$

4.1.4 Truncation Error

The following example specifies the truncation error made when using finite differences. Consider the function:

$$f(x) = A \sin(2\pi x/\lambda) \tag{4.10}$$

where A is a constant amplitude and λ is a certain wavelength. The derivative of this function is given by:

$$\frac{df}{dx} = 2\pi A/\lambda \cos(2\pi x/\lambda) \tag{4.11}$$

If we use the centred difference as a proxy for the first derivative, we obtain:

$$\frac{f(x+\Delta x) - f(x-\Delta x)}{2\Delta x} = \frac{A \sin[2\pi(x+\Delta x)/\lambda] - A \sin[2\pi(x-\Delta x)/\lambda]}{2\Delta x}$$

With some mathematical manipulation, the latter equation can be formulated as:

$$\frac{f(x+\Delta x) - f(x-\Delta x)}{2\Delta x} = 2\pi A/\lambda \cos(2\pi x/\lambda) \cdot [1-\epsilon]$$

where the relative error with respect to the true solution – *the truncation error* – is given by:

$$\epsilon(\Delta x) = 1 - \frac{\sin(2\pi \Delta x/\lambda)}{2\pi \Delta x/\lambda}$$

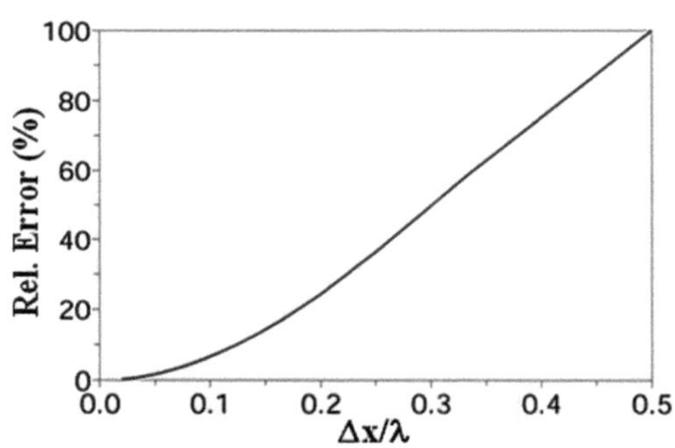

Fig. 4.2 Relative error (%) inherent with use of the centred scheme for (4.11) as a function of $\Delta x/\lambda$

The truncation error becomes reasonably small if we resolve the wavelength by more than 10 grid points (Fig. 4.2). In other words, when using finite differences, only waves with a wavelength greater than tenfold the grid spacing are resolved accurately. Similar conclusion can be drawn for time step requirements to resolve a given wave period.

4.2 Long Surface Gravity Waves

4.2.1 Extraction of Individual Processes

The Navier–Stokes equations describe a great variety of processes that can occur simultaneously in fluids on different time scales and lengthscales. Nevertheless, under certain assumptions, we can extract individual processes from these equations to study them in isolation from other processes. For instance, waves of a period short compared with the inertial period are unaffected by the Coriolis force and we can ignore the Coriolis force for such waves, which simplifies the governing equations. By making certain assumptions, we will progressively learn more about a variety of physical processes existing in fluids.

4.2.2 Shallow-Water Processes

From scaling considerations, it can be shown that the hydrostatic relation holds for processes of horizontal lengthscale exceeding by far the vertical lengthscale. These processes are referred to as *shallow-water processes*, even if they occur in the atmosphere or in deep portions of the ocean. It is the ratio between horizontal and vertical lengthscales that matters here!

4.2.3 The Shallow-Water Model

We consider a fluid layer of uniform density with a freely moving surface to study long surface waves of a wavelength long compared with the fluid depth (Fig. 4.3). We assume wave periods short compared with the inertial period, so that the Coriolis force can be neglected, and we simply ignore frictional effects to first-order approximation. We neglect the nonlinear terms (advection of momentum), which implies that the phase speed of waves exceeds by far the speed of water parcels. As another simplification, we consider waves that propagate exclusively along a channel aligned with the x-direction and being void of variations in the y-direction.

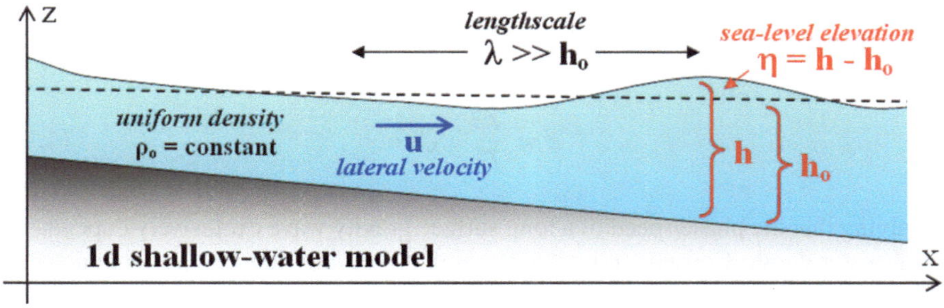

Fig. 4.3 Configuration of the one-dimensional shallow-water model. Undisturbed water depth is h_o

4.2.4 The Governing Equations

With the above simplifications, the equations governing the dynamics of long surface waves can be written as:

$$\frac{\partial u}{\partial t} = -g\frac{\partial \eta}{\partial x} \tag{4.12}$$

$$\frac{\partial \eta}{\partial t} = -\frac{\partial\,(u\;h)}{\partial x} \tag{4.13}$$

where u is speed in the x-direction, t is time, g is acceleration due to gravity, η is sea-level elevation, and h is total water depth.

The first equation is an expression of Newton's laws of motions and states that a slope in the sea surface operates to change the lateral velocity. The second equation – the vertically integrated form of the continuity equation – relates temporal changes in sea level to convergence/divergence of the depth-integrated lateral flow.

4.2.5 Analytical Wave Solution

Total water depth h can be approximated as constant for a flat seafloor together with wave amplitudes small compared with total water depth. In this case, the wave solution of the above equations is:

$$\eta(t,x) = \eta_o\ \sin\left(2\pi\ x/\lambda - 2\pi\ t/T\right) \tag{4.14}$$

$$u(t,x) = u_o\ \sin\left(2\pi\ x/\lambda - 2\pi\ t/T\right) \tag{4.15}$$

where η_o is wave amplitude, λ is wavelength, T is wave period, and the magnitude of u is given by:

$$u_o = \eta_o\sqrt{\frac{g}{h}}$$

It can also be shown (Cushman-Roisin, 1994) that these waves are governed by the well-known dispersion relation:

$$c = \frac{\lambda}{T} = \sqrt{gh} \tag{4.16}$$

implying that the phase speed of a long surface gravity wave exclusively depends on total water depth. Consequently, it follows that the ratio between horizontal speed of a water parcel and phase speed is very small:

$$\frac{u_o}{c} = \frac{\eta}{h} << 1$$

which is justification for neglection of the nonlinear terms. For instance, a long wave of 1 m in amplitude in a 100 m deep ocean propagates with a phase speed of about $c = 30$ m/s, while water parcels attain maximum lateral displacement speeds of only $u_o = 0.3$ m/s.

Horizontal flow under a long surface wave is depth-independent and so are horizontal gradients of u. On the basis of the local form of the continuity equation, given for our channel by:

$$\frac{\partial w}{\partial z} = -\frac{\partial u}{\partial x}$$

we can derive the solution for vertical speed of a fluid parcel as a function of depth:

$$w(t, x, z^*) = -2\pi \; u_o \; z^*/\lambda \; \cos(2\pi \; x/\lambda - 2\pi \; t/T)$$

where z^* is (positive) distance from the seafloor. Vertical speed vanishes at the plane seafloor (per definition) and approaches an oscillating maximum at the sea surface. The ratio between vertical and horizontal speeds of water parcels is $2\pi h/\lambda$. This ratio is small compared with unity for shallow-water waves ($\lambda >> h$). Accordingly, motions of water parcels in a shallow-water wave are largely horizontal. Another important feature inherent with long waves is that they reach the seafloor and are capable of stirring up sediment from the seafloor, if energetic enough. Figure 4.4 shows a snapshot of the analytical solution of a shallow-water wave.

4.2.6 Animation Script

A SciLab script, called "AnalWaveSol.sce", can be found in the folder "Miscellaneous/Waves" on the CD-ROM accompanying this book. This script creates an animation of the analytical wave solution.

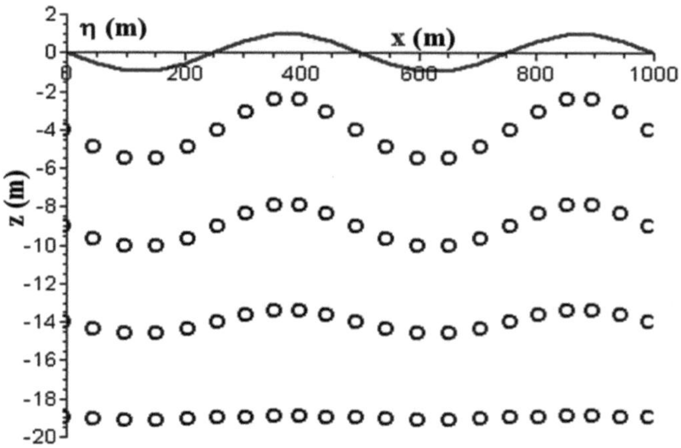

Fig. 4.4 Snapshot of a shallow-water wave of 1 m in amplitude and 500 m in wavelength in a 20 m deep ocean. Shown are sea-level elevation and vertical displacements of fluid parcels at selected depths

4.2.7 Numerical Grid

We use a spatial grid of constant grid spacing in which velocity grid points are located halfway between adjacent sea-level grid points (Fig. 4.5).

Fig. 4.5 The staggered grid. The cell index k refers to a certain grid cell. Water depth is calculated at sea-level grid points

4.2.8 Finite-Difference Scheme

On the basis of the staggered grid (see Fig. 4.5), the momentum equation (4.12) can be written in finite-difference form as:

$$u_k^{n+1} = u_k^n - \Delta t\ g\left(\eta_{k+1}^n - \eta_k^n\right)/\Delta x \tag{4.17}$$

where n is time level, k is grid index, Δt is time step, and Δx is grid spacing. A *control volume* (Fig. 4.6) is used to discretise equation (4.3). Accordingly, we can write this equation as:

$$\eta_k^* = \eta_k^n - \Delta t\left(u_k^{n+1}\ h_e - u_{k-1}^{n+1}\ h_w\right)/\Delta x \tag{4.18}$$

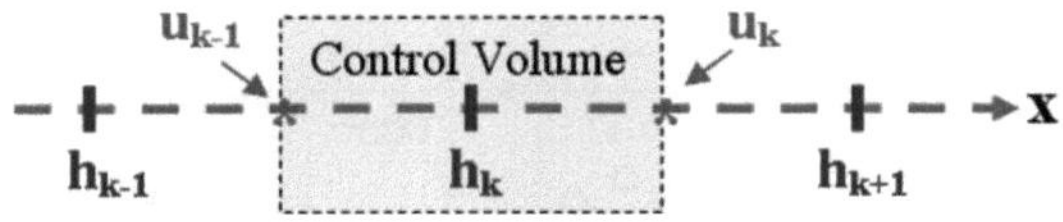

Fig. 4.6 The control-volume approach. A control volume of length Δx is centred around a water-depth grid point h_k. Temporal sea-level changes are computed from volume fluxes through the left and right-hand faces of the control volume using the upstream approach

where h_w and h_e, respectively, are the layer thicknesses at the western and eastern faces of the control volume. Input to this equation are prognostic values of u calculated a step earlier from (4.17). The final prediction for η will be slightly smoothed by applying a filter (see below) to η^*.

Here, the choices for h_w and h_e are made dependent of the flow direction at the respected face in an upstream sense. For example, we take $h_w = h^n_{k-1}$ for $u^{n+1}_{k-1} > 0$, but $h_w = h^n_k$ for $u^{n+1}_{k-1} < 0$. This can be elegantly formulated by means of:

$$\eta^*_k = \eta^n_k - \Delta t/\Delta x \left(u^+_k\, h^n_k + u^-_k\, h^n_{k+1} - u^+_{k-1}\, h^n_{k-1} - u^-_{k-1}\, h^n_k\right) \tag{4.19}$$

where

$$u^+_k = 0.5\left(u^{n+1}_k + \left|u^{n+1}_k\right|\right) \text{ and } u^-_k = 0.5\left(u^{n+1}_k - \left|u^{n+1}_k\right|\right)$$

This control-volume approach is numerically diffusive, but conserves volume of the water column.

4.2.9 Stability Criterion

The stability criterion for the above equations, known as Courant-Friedrichs-Lewy condition or *CFL condition* (Courant et al., 1928), is:

$$\lambda = \frac{\Delta t}{\Delta x}\sqrt{g\, h_{max}} \le 1 \tag{4.20}$$

where $h_{\max}$ is the maximum water depth encountered in the model domain. In other words, the time step is limited by:

$$\Delta t \le \frac{\Delta x}{\sqrt{g\, h_{\max}}}$$

which can be a problem for deep-ocean applications if a fine lateral grid spacing is required.

4.2.10 First-Order Shapiro Filter

As will be shown below, the finite-difference equation presented above are subject to oscillations developing on wavelengths of $2\Delta x$. Some of these oscillations might represent true physics, others might be artificial *numerical waves*. To remove these small-scale oscillations, the following first-order Shapiro filter (Shapiro, 1970) can be used:

$$\eta_k^{n+1} = (1-\epsilon)\eta_k^* + 0.5\epsilon(\eta_{k-1}^* + \eta_{k+1}^*) \tag{4.21}$$

where η_k^* are predicted from (4.19) and ϵ is a smoothing parameter. This method removes curvatures in distributions to a certain degree. The smoothing parameter in this scheme should be chosen as small as possible.

4.2.11 Land and Coastlines

Land grid points are realised by requesting absence of flow on land. In addition to this, no flow is allowed across coastlines unless a special flooding algorithm is implemented (see Sect. 4.4). The layer thickness h can be used as a control as to whether grid cells are "dry" or "wet" . Then, we can set u_k to zero in grid cells where $h_k \le 0$. Owing to the staggered nature of the grid (see Fig. 4.5), coastlines require the additional condition that u_k has to be zero if $h_{k+1} \le 0$.

4.2.12 Lateral Boundary Conditions

The model domain is defined such that the prediction ranges from $k = 1$ to $k = nx$. Values have to be allocated to the first and last grid cells of the model domain; that is, to $k = 0$ and $k = nx+1$ (Fig. 4.7). One option is to treat these boundaries as closed. Advective lateral fluxes of any property are eliminated via the statements:

$$u_0^n = 0$$
$$u_{nx}^n = 0$$

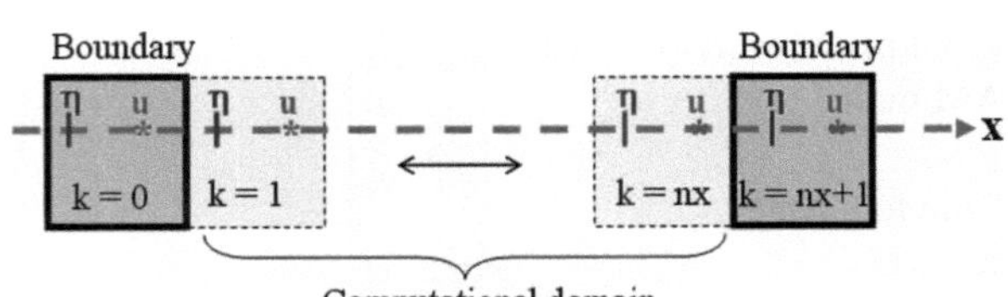

Fig. 4.7 The boundary grid cells of the model domain used for implementation of lateral boundary conditions

Zero-gradient boundary conditions are used to eliminate lateral diffusive fluxes of a property C. These conditions read:

$$C_0^n = C_1^n$$
$$C_{nx+1}^n = C_{nx}^n$$

Some applications justify the use of so-called *cyclic* boundary conditions that for a property C read:

$$C_0^n = C_{nx}^n$$
$$C_{nx+1}^n = C_1^n$$

Here, both ends of the model domain are connected to form a channel of infinite length.

4.2.13 Modular FORTRAN Scripting

It makes sense to split longer FORTRAN codes into several files containing different parts of the code containing the main code and separate *modules*. An example of a main code and two modules is given in the following. In this example, the main code is linked with two external modules via the command "USE".

The first module contains declarations of parameters and variables. The second module includes two subroutines. The statement CONTAINS is used if a module contains more than one subroutine or function. The first subroutine allocates initial values to parameters and the second one does a simple calculation. Subroutines are called with a CALL statement. The main code is given by:

```
PROGRAM main
USE decla
USE calcu
CALL init
CALL squaresum
z = z+c ! final calculation
write(6,*)"x = ",x ! print result on screen
END PROGRAM main
```

The "decla" module reads:

```
MODULE decla
REAL, PARAMETER :: c = 1.0
REAL :: x,y,z
```

```
END MODULE decla
```

The module "calcu" reads:

```
MODULE calcu
USE decla
CONTAINS
!+++++++++++++++++++++++++++++++++++++++++++
SUBROUTINE init
x = 2.0
y = 1.0
RETURN
END SUBROUTINE init
!+++++++++++++++++++++++++++++++++++++++++++
SUBROUTINE squaresum
real :: sum ! declaration of local variable
sum = x+y
z = sum*sum
RETURN
END SUBROUTINE squaresum
END MODULE calcu
```

As the reader learns from this example, modules can be used to share parameters, variables and subroutines. With sound use of modules, subroutines, and functions, FORTRAN codes can attain a much clearer structure.

4.2.14 Structure of the Following FORTRAN Codes

FORTRAN codes of the following exercises consist of three files: a main code, one module for declarations and another module comprising subroutines (Fig. 4.8). Compiling a FORTRAN code that contains modules consists of two steps. The modules are compiled first with:

```
g95 -c file2.f95 file3.f95
```

Then, the module files can be linked with the main code via:

```
g95 -o run.exe file1.f95 file2.o file3.o
```

The code can then be executed by entering "run.exe" in the Command Prompt window.

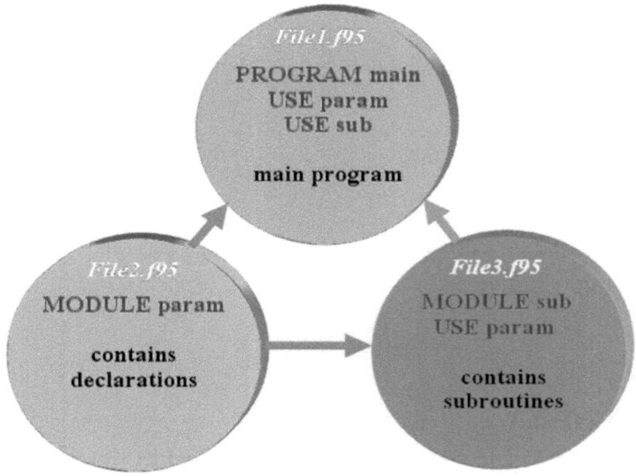

Fig. 4.8 Structure of the FORTRAN code for the following exercises

4.3 Exercise 5: Long Waves in a Channel

4.3.1 Aim

The aim of this exercise is to simulate the progression of shallow-water surface gravity waves in a channel of uniform water depth.

4.3.2 Instructions

We employ the finite-difference Eqs. (4.17), (4.19) and (4.21) for a one-dimensional channel with closed ends under a variety of initial and forcing conditions. In the following applications, the channel has a length of 1000 m being resolved by a grid spacing of $\Delta x = 10$ m. We choose 101 grid cells in the x-direction plus another grid cell on each end of the channel as boundary grid points. With the choice of an uneven number of grid points, the centre of the channel is defined by a single grid cell.

Undisturbed water depth is set to 10 m. Dry boundary cells are assigned a water depth of zero. Two different forcing scenarios are considered (Fig. 4.9). In Scenario 1, we commence the simulation with a 110-m wide region centred in the channel in which sea level is initially 1 m higher than elsewhere. This scenario is referred to as "dam-break simulation".

In Scenario 2, we place a wave paddle in the middle of the channel and let the sea level oscillate with an amplitude of 1 m and a period of 20 s. In both scenarios, the solutions are explored for different values of the parameter ϵ in the Shapiro filter. The choice of $\epsilon = 0$ switches off this filter. The time step is set to $\Delta t = 0.1$ s, which satisfies the CFL stability criterion.

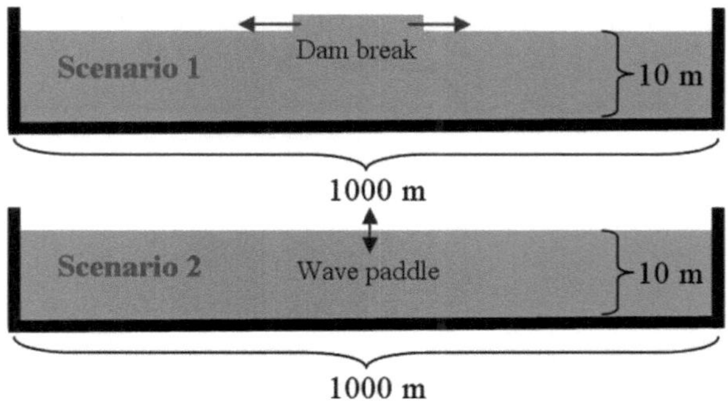

Fig. 4.9 Two different scenarios considered in Exercise 5

4.3.3 Sample Code and Animation Script

The folder "Exercise 5" on the CD-ROM contains the computer codes for this exercise. The file "info.txt" contains additional information.

4.3.4 Results

The initial sea-level anomaly, prescribed in the dam-break simulation, disintegrates into two separate wave bulges propagating toward opposite ends of the channel. Each of these waves is reflected at the respective boundary, and both waves meet again after some time in the centre of the channel. Without use of the Shapiro filter ($\epsilon = 0$), unwanted irregularities appear in sea-level elevations and, consequently, in the flow field (Fig. 4.10).

With the choice of $\epsilon = 0.05$, the output fields are much smoother (Fig. 4.11), but, apparently, the Shapiro filter has introduced some *numerical diffusion*. The reader is encouraged to "play around" with different values of ϵ until the result looks "reasonable". Wave paddle forcing (Scenario 2) creates surface gravity waves that propagate toward both ends of the channel (not shown). Reflection at boundaries triggers wave interference.

4.4 Exercise 6: The Flooding Algorithm

4.4.1 Aim

We want to be able to simulate transient flooding of dry regions (land) of the model domain. There are several applications of this, such as transient flooding of coastlines caused by tides or tsunamis. This requires special numerical treatment not included in the previous code.

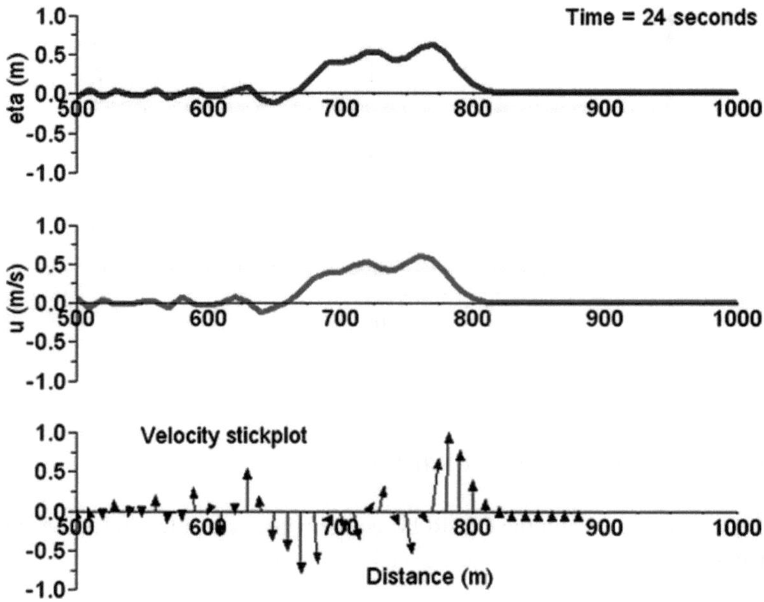

Fig. 4.10 Exercise 5. Scenario 1. Model output after 24 s of simulation for the right half of the channel for $\epsilon = 0$. The *top panel* shows sea-level elevation (m), the *middle panel* horizontal velocity (m/s), and the *bottom panel* velocity vectors (stickplots). Stickplots are strongly deformed owing to the axis ratio chosen

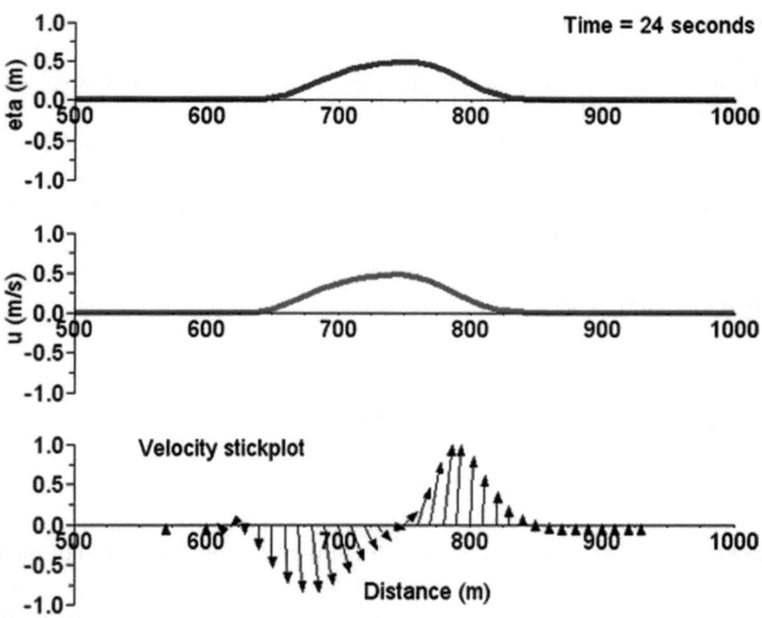

Fig. 4.11 Exercise 5. Scenario 1. Same as Fig. 4.9, but with $\epsilon = 0.05$

4.4.2 Redefinition of Wet and Dry

From now on, wet grid cells are defined as cells of total water depths h exceeding a certain threshold value, h_{min}, typically set to a few centimetres. This threshold is required to avoid that fluid retreating from a wet region produces negative layer thicknesses that will cause the model to crash. Accordingly, dry grid cells are defined as cells where $h \leq h_{\mathrm{min}}$.

4.4.3 Enabling Flooding of Dry Grid Cells

Flooding of dry grid cells is implemented in the code via calculation of the horizontal speed at the interface between wet and dry grid cells. This calculation is performed whenever the pressure-gradient force is directed toward the dry cell. Otherwise, velocity at this grid point is kept at zero value. With a nonzero inflow, the water level in the dry cell will rise and this cells eventually turns into a wet grid cell once the layer thickness exceeds h_{min}.

4.4.4 Flooding of Sloping Beaches

The pressure-gradient force is evaluated from the slope of the sea-level elevation with respect to an undisturbed state. Flooding of a sloping beach has to be treated the same way. Here, the sea-level elevation has to be defined as the distance of the sea surface from the undisturbed sea level, and not from the beach surface. Implementation of this flooding is done with the following steps, with reference to the illustration shown in Fig. 4.12.

1. Starting point is a certain distribution of bathymetry h_o with positive values referring to the ocean and negative values referring to elevated land surfaces.
2. Initial sea-level elevations η are assigned zero values in ocean regions, but follow land elevations with positive values there.

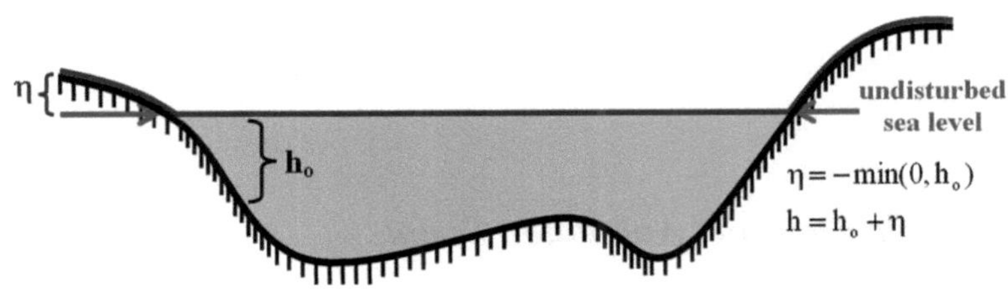

Fig. 4.12 Definitions for the flooding algorithm. Bathymetry (h_o) refers to (positive) total water depth in "wet" regions and (negative) land elevation in "dry" regions of the model domain. Elevation (η) is either sea-level elevation or land elevation. True layer thickness is the sum of h_o and η

3. True layer thickness is defined as the sum $h = h_o + \eta$. This true layer thickness is used for the classification of wet/dry grid cells on the basis of a threshold layer thickness $h_{\min}$.

4.4.5 Ultimate Crash Tests

The following scenarios can be regarded as ultimate crash tests of the flooding algorithm (Fig. 4.13). Scenario 1 considers a tsunami wave spilling over an island whose tip is located just 0.5 m above undisturbed sea level. The tsunami is created by prescription of region, 200 m in width, in which sea level is initially elevated by 1 m. Lateral boundaries are closed. Will the model be able to describe the dynamics of this tsunami as it spills over the tip of the island?

The second scenario is even more extreme. This scenario considers a hillside of a certain slope and including a depression. The task is to simulate the dynamics of a plume of water, initially 1 m thick and 200 m in horizontal extent, cascading downward on the hillside. Zero-gradient conditions are imposed for all variables at lateral boundaries. Will the model be able to cope with this task? After passage of the plume, will there be water left over in the depression?

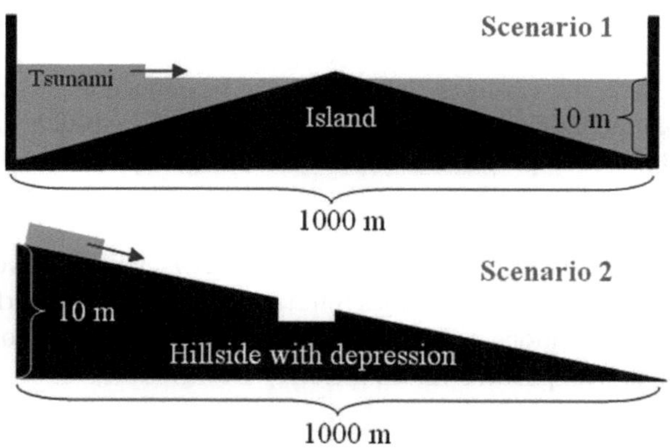

Fig. 4.13 Exercise 6. Ultimate crash tests of the flooding algorithm

4.4.6 Sample Code and Animation Script

FORTRAN simulation codes and Scilab animation scripts for both scenarios can be found in the folder "Exercise 6" of the CD-ROM. Several changes of the FORTRAN code for Exercise 5 are required for implementation of the flooding algorithm. The file "info.txt" outlines these changes.

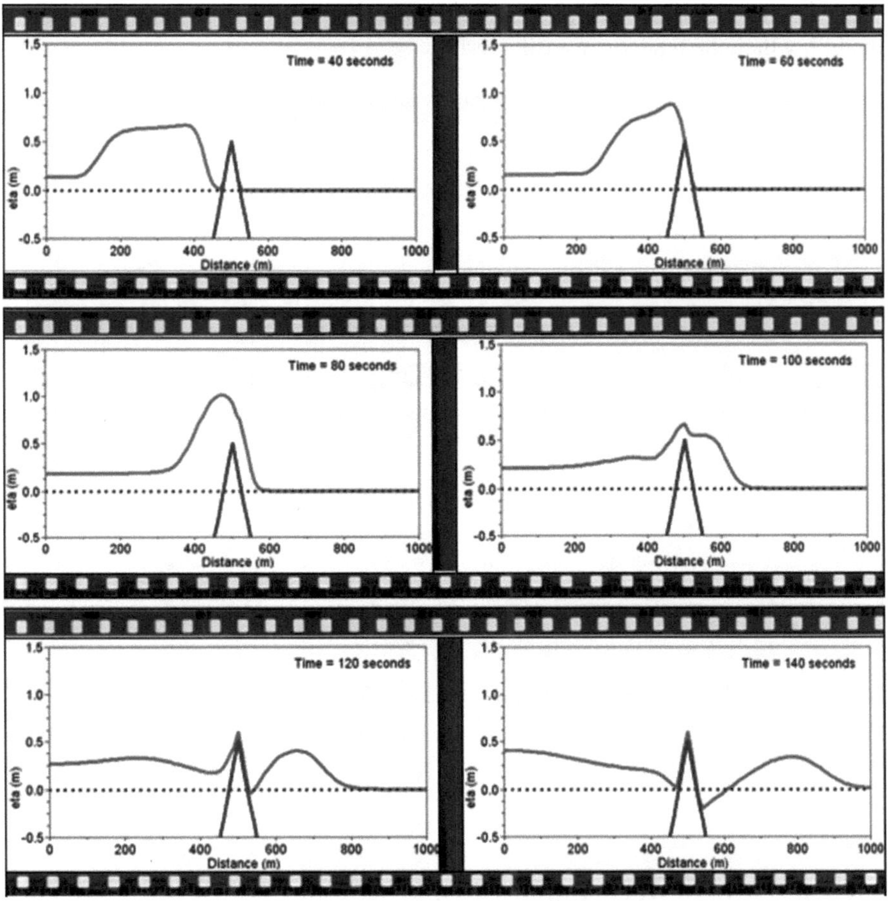

Fig. 4.14 Exercise 6. Scenario 1. Snapshots of sea-level elevation after 80, 100, 120 and 140 s of simulation. The *dotted line* indicates the location of undisturbed sea level

4.4.7 Results

The model can successfully simulate tsunami-caused flooding of an island (Fig. 4.14). A couple of features should be highlighted. First, the amplitude of the wave increases as it approaches shallower water. This happens in a similar fashion for wind waves approaching a beach. Secondly, only a fraction of water passes the island creating a smaller tsunami in the lee of the island. A fraction of water is blocked by the island and becomes reflected to form a tsunami propagating into the opposite direction. In the end, the island is void of water again, except for a thin inactive layer of thickness $h_{\min}$.

The model is also capable of simulating a plume cascading avalanche-like downward on a hillside and filling a depression in the ground (Fig. 4.15). All this takes place well above sea level! The plume dynamics are influenced by the steepness of the hill, which the reader can easily verify.

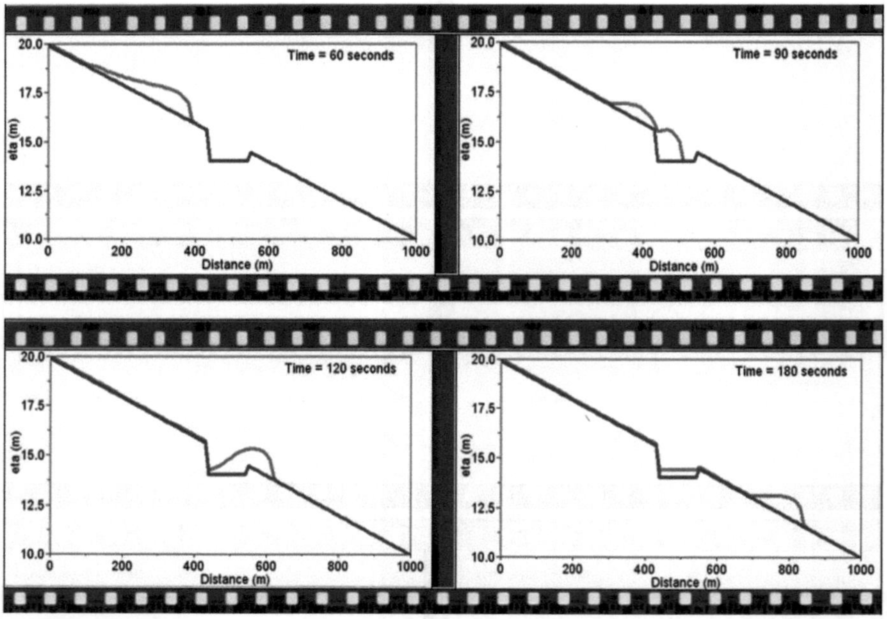

Fig. 4.15 Exercise 6. Scenario 2. Snapshots of sea-level elevation after 60, 90, 120 and 180 s of simulation

4.5 The Multi-Layer Shallow-Water Model

4.5.1 Basics

Under the shallow-water assumption, we can formulate a dynamical model for a water column consisting of multiple layers each of a certain density. Figure 4.16 gives the configuration of a multi-layer model. Layers are numbered from top to bottom and the surface layer carries the index 1. The momentum equation for a layer i is given by:

$$\frac{\partial u_i}{\partial t} = -\frac{1}{\rho_i}\frac{\partial P_i}{\partial x} \tag{4.22}$$

where i is the layer index. Dynamic pressure in the layers is given by:

$$P_1 = \rho_1 \, g \, \eta_1$$
$$P_2 = P_1 + (\rho_2 - \rho_1) \, g \, \eta_2$$

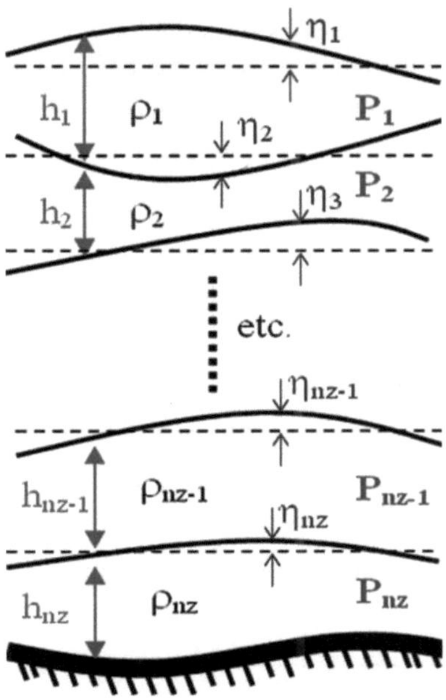

Fig. 4.16 Configuration of a multi-layer shallow-water model. The total number of layers is nz

$$
\begin{aligned}
P_3 &= P_2 + (\rho_3 - \rho_2)\, g\, \eta_3 \\
&\vdots \\
P_{nz-1} &= P_{nz-2} + (\rho_{nz-1} - \rho_{nz-2})\, g\, \eta_{nz-1} \\
P_{nz} &= P_{nz-1} + (\rho_{nz} - \rho_{nz-1})\, g\, \eta_{nz}
\end{aligned}
$$

which can be written in generalised form as:

$$P_i = P_{i-1} - (\rho_i - \rho_{i-1})\, g\, \eta_i \quad \text{for} \quad i = 1, 2, 3, \ldots, nz \tag{4.23}$$

where i is the layer index and η_i are interface displacements with reference to certain equilibrium levels. The latter equations require iteration from top to bottom with the boundary setting $P_0 = 0$ and $\rho_0 = 0$, which disables atmospheric pressure and sets air density to zero.

Conservation of volume in each layer corresponds to the prognostic equations for layer-thickness:

$$\frac{\partial h_i}{\partial t} = -\frac{\partial\,(u_i\, h_i)}{\partial x} \tag{4.24}$$

Layer thicknesses are given by:

$$h_i = h_{i,o} + \eta_i - \eta_{i+1} \tag{4.25}$$

where $h_{i,o}$ are undisturbed thicknesses for a fluid at rest and $\eta_{nz+1} = 0$ represents a rigid seafloor. With use of the latter relations, the layer-thickness equations turn into prognostic equations for interface displacements, given by:

$$\begin{aligned}
\frac{\partial \eta_{nz}}{\partial t} &= -\frac{\partial (u_{nz}\, h_{nz})}{\partial x} \\
\frac{\partial \eta_{nz-1}}{\partial t} &= -\frac{\partial (u_{nz-1}\, h_{nz-1})}{\partial x} + \frac{\partial \eta_{nz}}{\partial t} \\
&\vdots \\
\frac{\partial \eta_{2}}{\partial t} &= -\frac{\partial (u_{2}\, h_{2})}{\partial x} + \frac{\partial \eta_{3}}{\partial t} \\
\frac{\partial \eta_{1}}{\partial t} &= -\frac{\partial (u_{1}\, h_{1})}{\partial x} + \frac{\partial \eta_{2}}{\partial t}
\end{aligned}$$

These equations can be written in the generalised form:

$$\frac{\partial \eta_i}{\partial t} = -\frac{\partial (u_i\, h_i)}{\partial x} + \frac{\partial \eta_{i+1}}{\partial t} \quad \text{for} \quad i = nz, nz-1, \ldots, 1 \tag{4.26}$$

with $\eta_{nz+1} = 0$ representing the rigid seafloor. Note that, in contrast to the pressure iteration, this interaction goes from bottom to top.

4.6 Exercise 7: Long Waves in a Layered Fluid

4.6.1 Aim

The aim of this exercise is to simulate the progression of long gravity waves in a fluid consisting of multiple layers of different densities.

4.6.2 Task Description

We consider a stratified fluid consisting of ten layers of an initial thickness of 10 m each. The density of the top layer is 1025 kg/m^3 and density increases from 1026 kg/m^3 to 1026.5 kg/m^3 from the second layer to the bottom layer. In addition to this, we add a simple bathymetry including a riff (Fig. 4.17) to test the multi-layer flooding algorithm. The model is forced by prescribing sinusoidal oscillations of surface and interface displacements of an amplitude of 1m near the western boundary. Lateral boundaries are closed.

Two different forcing periods are considered. The first experiment uses a forcing period of 10 s. The forcing period in the second experiment is 2 h. Simulations run over 10 times the respective forcing period and data outputs are produced at intervals of a tenth of the forcing period. The time step is set to $\Delta t = 0.25$ s in both cases.

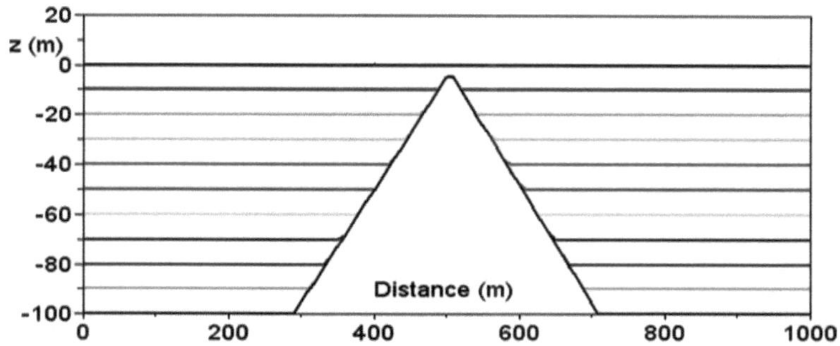

Fig. 4.17 Configuration for Exercise 7

4.6.3 Sample Code and Animation Script

The folder "Exercise 7" on the CD-ROM contains the computer codes for this exercise. The "info.txt" gives additional information.

4.6.4 Results

A relatively short forcing period of 10 s creates barotropic surface gravity waves (Fig. 4.18). Density interfaces oscillate in unison with the sea surface. The phase speed of the wave is $c \approx \sqrt{gH}$, where H is total depth of the water column. Waves approaching the riff pile up while their wavelength decreases. The wave therefore becomes steeper aiming to break. Indeed, wave breaking cannot be simulated with a layer model. Notice that waves continue to propagate eastward on the lee side of the riff.

When watching the real sea patiently, riffs or sandbars can be identified as the regions with locally increased wave heights and wave breaking. This process is known as *wave shoaling*.

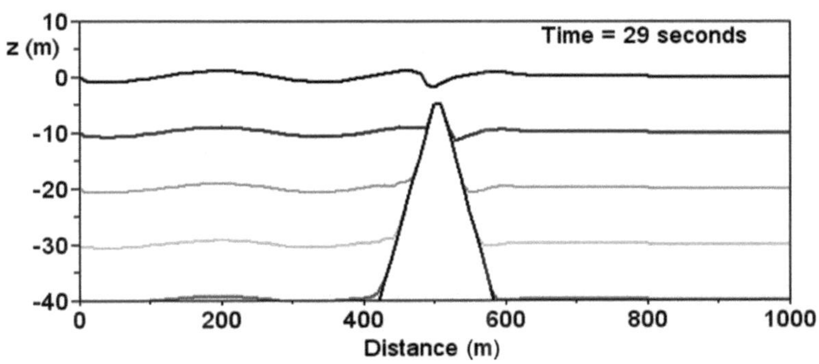

Fig. 4.18 Exercise 7. Snapshot of surface and interface displacements for a forcing period of 10 s. Only the top 40 m of the water column are shown

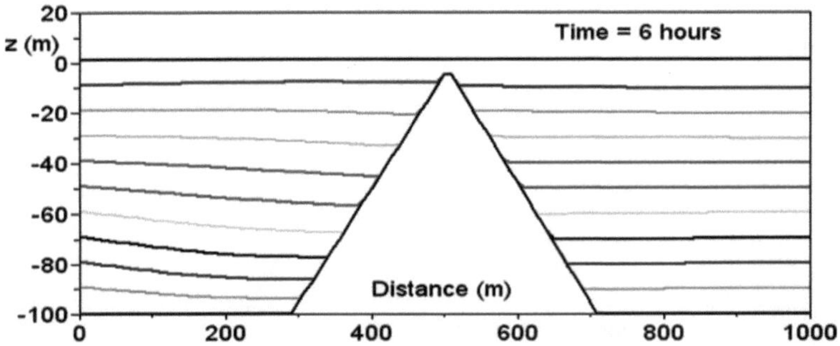

Fig. 4.19 Exercise 7. Snapshot of surface and interface displacements for a forcing period of 2 h

In contrast to this, a longer forcing period of 2 h creates internal waves, largely blocked by the riff, of wave heights >10 m (Fig. 4.19). Although all layers oscillate the same way near the forcing location, interfaces oscillate in a complex stretching and shrinking pattern near the riff.

4.6.5 Phase Speed of Long Internal Waves

In a two-layer fluid, it can be shown that the phase speed of long interfacial waves is given by (see Pond and Pickard, 1983):

$$c_{iw} = \sqrt{g'h^*} \tag{4.27}$$

where g' is reduced gravity, and $h^* = h_1h_2/(h_1 + h_2)$ is a reduced depth scale with h_1 and h_2 being the undisturbed thicknesses of the top and bottom layers, respectively. For $h_2 >> h_1$, we yield $h^* \approx h_1$. Internal gravity waves propagate much slower compared with surface gravity waves. Their periods and amplitudes are much greater and, like surface waves, internal waves can break under certain conditions. Indeed, breaking of internal waves cannot by simulated with a hydrostatic layer model.

4.6.6 Natural Oscillations in Closed Bodies of Fluid

Closed water bodies such as a lake or a fish tank are subtle to natural oscillations. *Wave nodes* are the locations at which the fluid only experiences horizontal but no vertical motions. In contrast to this, *anti-nodes* are locations that experience maximum vertical motion, but no or only little horizontal motions. Natural oscillations are standing waves (phase speed is virtually zero) that display anti-nodes of maximum vertical displacements of the surface (or density interfaces) at the ends of the basin.

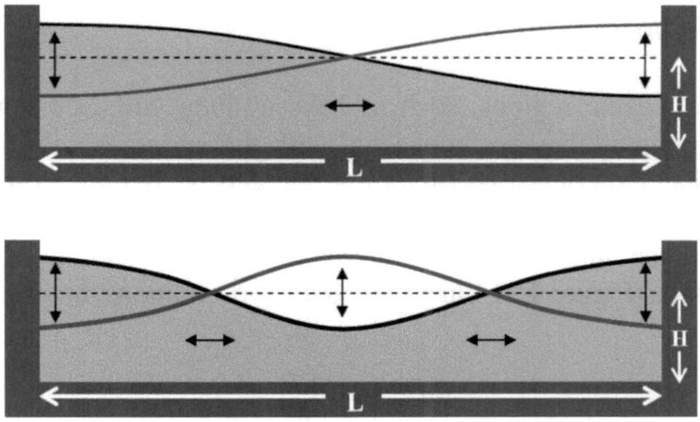

Fig. 4.20 Examples of natural oscillations that occur in a closed channel

For example, consider a long, shallow and closed channel of constant water depth H and length L. The basic natural oscillation consists of a single node in the channel's centre and anti-nodes at both ends (Fig. 4.20). The next higher-order natural oscillation is one with two nodes in the channel, the next one comes with three nodes, and so on. A systematic analysis reveals that natural oscillations occur for $L = m/2\lambda$, where $m = 1, 2, 3, \ldots$ is the number of wave nodes establishing in the channel, and λ is wavelength. In general forms, we can write the latter *resonance conditions* as:

$$T = \frac{\lambda}{c} = \frac{2}{m}\frac{L}{c} \quad \text{for} \quad m = 1, 2, 3, \ldots \tag{4.28}$$

where T is wave period (or forcing period of a wave paddle) and c is the phase speed of waves, which can be either surface or interfacial waves.

4.6.7 Merian's Formula

With the dispersion relation for long surface gravity waves, $c = \sqrt{gH}$, forcing periods triggering a so-called *resonance response* in a closed channel, are given by:

$$T \approx \frac{2}{m}\frac{L}{\sqrt{gH}} \quad for \quad m = 1, 2, 3, \ldots \tag{4.29}$$

This is known as *Merian's formula* (Merian, 1828). Resonance of internal waves occurs the same way, but for much longer forcing periods (since the phase speed of internal waves is much smaller compared with surface gravity waves).

4.6.8 Co-oscillations in Bays

Semi-enclosed oceanic regions, such as bays or long gulfs, do not exhibit free natural oscillations as they are connected with the ambient ocean. Nevertheless, these regions can experience *co-oscillations* forced by sea-level oscillations near the entrance. Co-oscillations of large amplitudes are exited if a wave node is located in vicinity of the entrance, such that water is pumped into and out of the bay in an oscillatory fashion. Figure 4.21 illustrates such co-oscillations that occur for forcing periods of:

$$T \approx \frac{\lambda}{c} \approx \frac{4}{m}\frac{L}{c} \quad \text{for} \quad m = 1, 3, 5, \ldots \tag{4.30}$$

where c is the phase speed of surface or internal gravity waves.

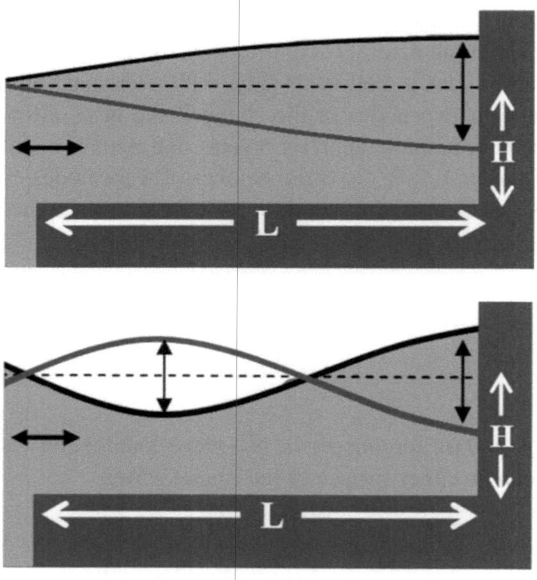

Fig. 4.21 Examples of co-oscillations in a semi-enclosed channel

4.6.9 Additional Exercise for the Reader

The task is to employ the shallow-water model with two layers to explore co-oscillations in a semi-enclosed bay. Figure 4.22 displays the physical settings. As in Exercise 5, forcing is provided by placing a wave paddle near the left end of the

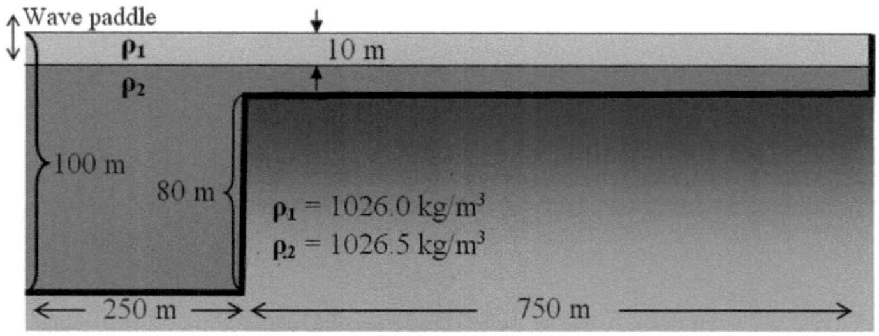

Fig. 4.22 Configuration of the additional exercise

model domain. The reader should try forcing periods of around 5.4 h for stimulation of large-amplitude internal co-oscillations in the bay.

Chapter 5
2D Shallow-Water Modelling

Abstract This chapter applies the two-dimensional shallow-water equations to study various processes such as surface gravity waves, the wind-driven circulation in a lake, the formation of turbulent island wakes, and the barotropic instability mechanism. The reader is introduced to various advection schemes simulating the movement of Eulerian tracer and describing the nonlinear terms in the momentum equations.

5.1 Long Waves in a Shallow Lake

5.1.1 The 2D Shallow-Water Wave Equations

We assume a lake of uniform water density and allow for variable bathymetry. For simplicity, frictional effects and the coriolis force are ignored and so are the nonlinear terms. This implies that our waves have a period short compared with the inertial period and that the phase speed of waves exceeds flow speeds by far. Under these assumptions, the momentum equations can be formulated as:

$$
\begin{aligned}
\frac{\partial u}{\partial t} &= -g\frac{\partial \eta}{\partial x} \\
\frac{\partial v}{\partial t} &= -g\frac{\partial \eta}{\partial y} \\
\frac{\partial \eta}{\partial t} &= -\frac{\partial (u\ h)}{\partial x} - \frac{\partial (v\ h)}{\partial y}
\end{aligned}
\tag{5.1}
$$

where u and v are components of horizontal velocity, t is time, g is acceleration due to gravity, η is sea-level elevation, and h is total water depth.

5.1.2 Arakawa C-grid

The *Arakawa C-grid* (Arakawa and Lamb, 1977) is a staggered numerical grid in which the components of velocity are found between adjacent sea-level grid points

J. Kämpf, *Ocean Modelling for Beginners*,
DOI 10.1007/978-3-642-00820-7_5,

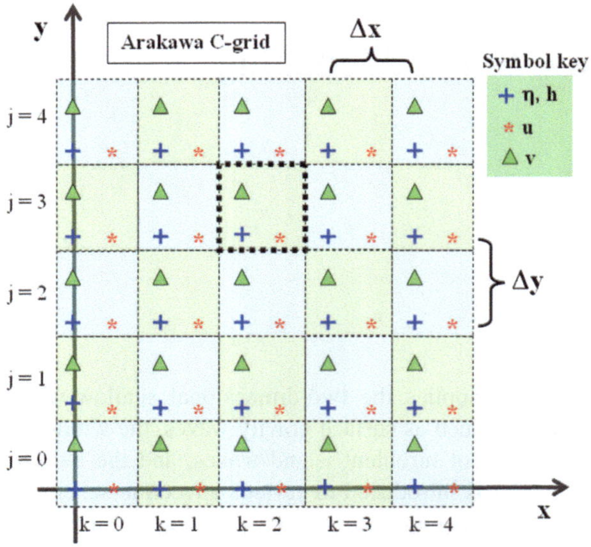

Fig. 5.1 Configuration of the horizontal version of the Arakawa C-grid. The grid cell with the grid index $j = 3$ and $k = 2$ is highlighted the doted rectangle highlights the grid cell with index

(Fig. 5.1). This grid, being widely used by the oceanographic modelling community, will be the basis of the following model codes. Note that u and v velocity components are not located at the same grid points.

5.1.3 Finite-Difference Equations

We need to have two cell indices in this two-dimensional application with j being the cell index in the y-direction and k being the cell index in the x-direction. With reference to the Arakawa C-grid, the governing equations can be written in finite-difference form as:

$$
\begin{aligned}
u_{j,k}^{n+1} &= u_{j,k}^{n} - \Delta t \; g \left(\eta_{j,k+1}^{n} - \eta_{j,k}^{n}\right) / \Delta x \\
v_{j,k}^{n+1} &= v_{j,k}^{n} - \Delta t \; g \left(\eta_{j+1,k}^{n} - \eta_{j,k}^{n}\right) / \Delta y \\
\eta_{j,k}^{*} &= \eta_{j,k}^{n} - \Delta t \left\{ \left(u_{j,k}^{n+1} \, h_e - u_{j,k-1}^{n+1} \, h_w\right) / \Delta x - \left(v_{j,k}^{n+1} \, h_n - v_{j-1,k}^{n+1} \, h_s\right) / \Delta y \right\}
\end{aligned}
\tag{5.2}
$$

where h_w and h_e are layer thicknesses at the western and eastern faces of the control volume, and h_s and h_n are layer thicknesses at the southern and northern faces of the control volume. Again, the upstream scheme is used to specify the grid indices used for of these thicknesses (see Sect. 4.2).

In addition to this, sea-level elevations are slightly smoothed with use of the two-dimensional version of the first-order Shapiro filter. To this end, values of sea-level elevation for the next time level follow from:

$$\eta_{j,k}^{n+1} = (1-\epsilon)\eta_{j,k}^{*} + 0.25\epsilon\left(\eta_{j,k-1}^{*} + \eta_{j,k+1}^{*} + \eta_{j-1,k}^{*} + \eta_{j+1,k}^{*}\right) \tag{5.3}$$

The parameter ϵ determines the degree of smoothing.

5.1.4 Inclusion of Land and Coastlines

As in the 1D shallow water model, water-depth values determine whether gridpoints belong to the ocean or to the land and also the location of coastlines. Land is here defined as zero or negative values of water depth and velocity components are kept at zero values in these "dry" grid cells during a simulation. Coastlines are implicitly defined by setting the flow component normal to a coastline to zero. Owing to the staggered nature of the Arakawa-C grid (see Fig. 5.1), this implies that u values are set to zero if there is land in the adjacent grid cell to the east. Accordingly, v values are set to zero in case of land in the adjacent grid cell to the north. Figure 5.2 gives an example of the shape of land and coastlines in the Arakawa-C grid. The flooding algorithm can be implemented in analog to the 1D application (see Sect. 4.4).

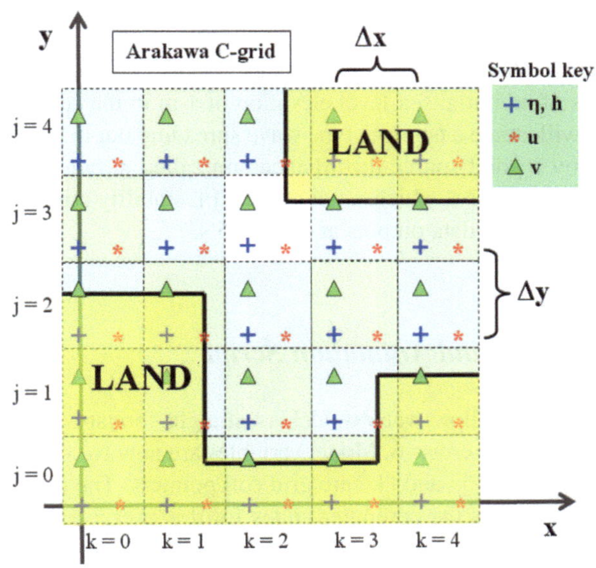

Fig. 5.2 Example of land and coastlines in the Arakawa C-grid

5.1.5 Stability Criterion

The CFL criterion for the two-dimensional shallow-water equations is given by:

$$\Delta t \leq \frac{\min(\Delta x, \Delta y)}{\sqrt{2gh_{\max}}} \tag{5.4}$$

where $h_{\max}$ is the maximum water depth encountered in the model domain.

5.2 Exercise 8: Long Waves in a Shallow Lake

5.2.1 Aim

The aim of this exercise is to simulate the progression of long circular surface gravity waves in a two-dimensional domain.

5.2.2 Task Description

We consider a square lake of 500 m× 500 m in areal extent and 10 m in depth using equidistant lateral grid spacings of $\Delta x = \Delta y = 10$ m. Lateral boundaries are closed. The flooding algorithm is included. Lake water is of uniform density. Forcing consists of an initial sea-level elevation of 1 m in the central grid cell that, when released, will create a tsunami-type wave spreading out in all directions. Such waves, created by a point-source disturbance, are called *circular waves*. The time step is chosen at $\Delta t = 0.1$ s, which satisfies the CFL stability criterion. The simulation is run for 100 s with data outputs at every 0.5 s.

5.2.3 Sample Code and Animation Script

The two-dimensional shallow-water model is a straight-forward extension of the 1D channel model used in Exercise 6. Model variables are now two-dimensional arrays such as "eta(j,k)" where "j" and "k" are grid cell pointers. The folder "Exercise 8" of the CD-ROM contains the computer codes for this exercise. The file "info.txt" contains additional information. Note that SciLab animation scripts can be run while the FORTRAN code is executed in the background. This is useful for long simulations to check whether the results are reasonable. If not, the FORTRAN run can be stopped by simultaneously pressing <Ctrl> and <c> in the Command Prompt window.

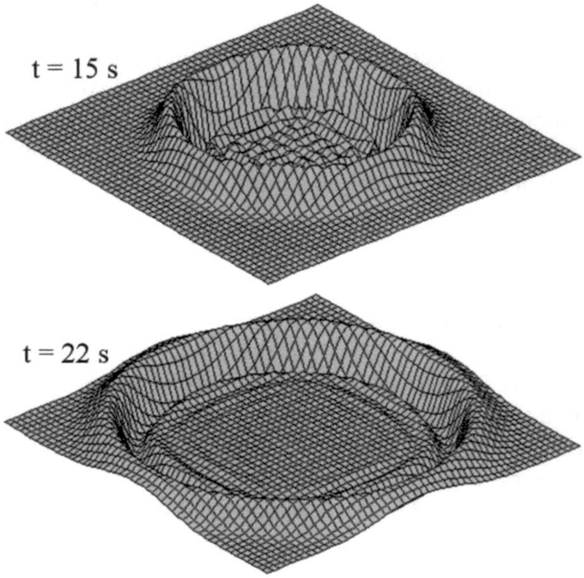

Fig. 5.3 Exercise 8. Sea-level elevations at selected times of the simulation

5.2.4 Snapshot Results

Figure 5.3 shows snapshot results of sea-surface elevation fields at selected times. As the reader can see, the model appears to be able to simulate the evolution and propagation of shallow-water waves in a two-dimensional domain.

5.2.5 Additional Exercise for the Reader

Add one or more islands or submerged seamounts to the model domain, and explore how long surface gravity waves deal with such obstacles.

5.3 Exercise 9: Wave Refraction

5.3.1 Aim

The aim of this exercise is to predict the dynamical behaviour of long, plane surface gravity waves as they approach shallower water in a coastal region.

5.3.2 Background

Why do plane surface gravity waves usually align their crests parallel to the beach as they approach the coast? The reason for this *wave refraction* is that all surface gravity waves eventually become long waves as they approach shallower water. The phase speed of long surface gravity waves depends exclusively on the total water depth. Portions of a wave located in deeper water travel faster than those in shallower water. Accordingly, the wave pattern experiences a gradual change of its orientation, such that wave crests become more and more aligned with topographic contours.

5.3.3 Task Description

The model domain is 2 km long and 500 m wide, resolved by grid spacings of $\Delta x = \Delta y = 10$ m (Fig. 5.4). The total water depth gradually decreases from 30 m at the western boundary to zero at the coast. The beach has a gentle slope of 10 cm per 10 m. Bathymetric contours and the coastline are rotated by 30° with respect to the y direction. A separate FORTRAN code is used to create this bathymetry as input for the simulation code.

Plane waves are waves whose wavefronts (crests and troughs) are straight and parallel to each other. Propagation occurs in a direction normal to wavefronts and can be described by means of a phase velocity vector. Such plane waves are generated at the western open boundary via prescription of sinusoidal sea-level oscillations (uniform along this boundary) of a period of 20 s.

In a water depth of 30 m, the forcing applied creates plane shallow-water waves of a wavelength ($\lambda = T\sqrt{gh}$) of approximately 340 m. The amplitude of oscillations is 20 cm. The northern and southern boundaries are open boundaries. The numerical time step is set to $\Delta t = 0.2$ s. The total simulation time is 200 s.

5.3.4 Lateral Boundary Conditions

If the prediction loop is performed from $j = 1$ to $j = ny$, the finite-difference equations (5.2) require a boundary condition for $\eta_{ny+1,k}$ at the northern open boundary and for $v_{0,k}$ at the southern open boundary (Fig. 5.5). To make these boundary conditions more consistent, $v_{0,k}$ can be included in the prediction, so that in analog to the northern boundary, a boundary condition for $\eta_{0,k}$ is now required. Note that

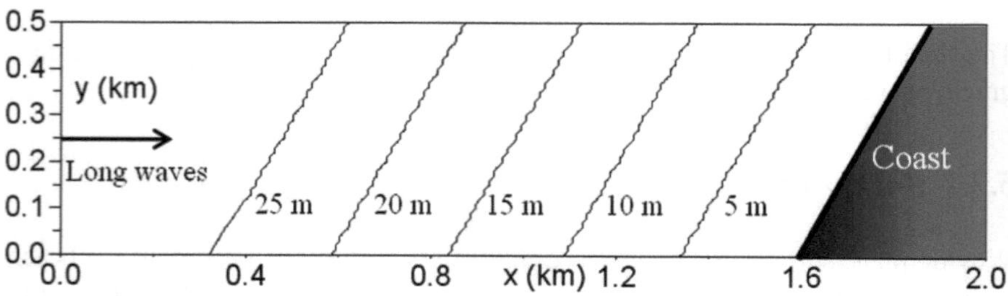

Fig. 5.4 Model configuration for Exercise 9

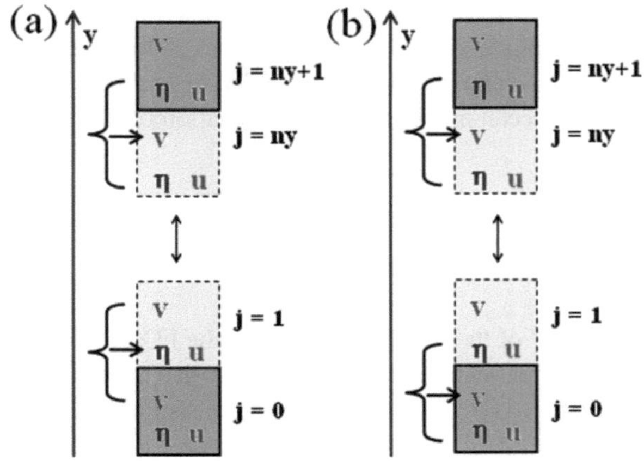

Fig. 5.5 Boundary points that need to be specified in the numerical code. In (**a**) v is predicted from $j = 1$ to $j = ny$, whereas in (**b**) the prediction loop includes $j = 0$

boundary values of u at the northern and southern open boundaries are not required since the finite-difference equations do not access these values.

In this model application, we include $j = 0$ in the prediction of v and assume that the second spatial derivative of η normal to the open boundary vanishes; that is:

$$\eta^n_{0,k} = 2\eta^n_{1,k} - \eta^n_{2,k}$$
$$\eta^n_{ny+1,k} = 2\eta^n_{ny,k} - \eta^n_{ny-1,k}$$

This condition induces a bias of the phase speed of waves in vicinity of the boundary and can lead to severe misrepresentation of the dynamics in certain applications. This condition works reasonably well for the configuration of this exercise. Advanced methods are available for the numerical treatment of waves approaching an open boundary such as *radiation conditions* first suggested by Sommerfeld (1949). The basis of radiation conditions is that the phase speed normal to an open boundary is the carrier of sea-level gradients across this boundary. This can be formulated by an advection equation reading:

$$\frac{\partial \eta}{\partial t} + c_y \frac{\partial \eta}{\partial y} = 0 \qquad (5.5)$$

where c_y is the phase speed normal to the boundary. For processes of variable phase speed, Orlanski (1976) proposed to compute c_y using a diagnostic version of (5.5); that is;

$$c_y = -\frac{\partial \eta / \partial t}{\partial \eta / \partial y}$$

where the partial derivatives are evaluated at previous time steps and from interior values. The computed phase speed is then used in (5.5) to predict variations of the boundary value of η. This numerical scheme, called *Orlanksi radiation condition*, is widespreadly used by ocean modellers. Implementation of this scheme remains for the advanced reader.

5.3.5 Sample Code and Animation Script

The folder "Exercise 9" of the CD-ROM contains the FORTRAN simulation code, a SciLab animation script, and a FORTRAN code called "topo.f95" that creates the bathymetry. The file "info.txt" gives additional information.

5.3.6 Results

The boundary forcing generates long surface gravity waves propagating into the model domain (Fig. 5.6). The wavelength becomes shorter in shallower water, which results in gradual alignments of the wave crests parallel to the coastline. Lateral boundary conditions do not lead to noticeable problems. The reason for wave steepening is that the phase speed under a wave crest exceed that under a wave trough according to (4.16).

5.3.7 Additional Exercise for the Reader

Repeat this exercise with forcing periods of 10 and 40 s. The advanced reader is encouraged to run this exercise with a modified bathymetry.

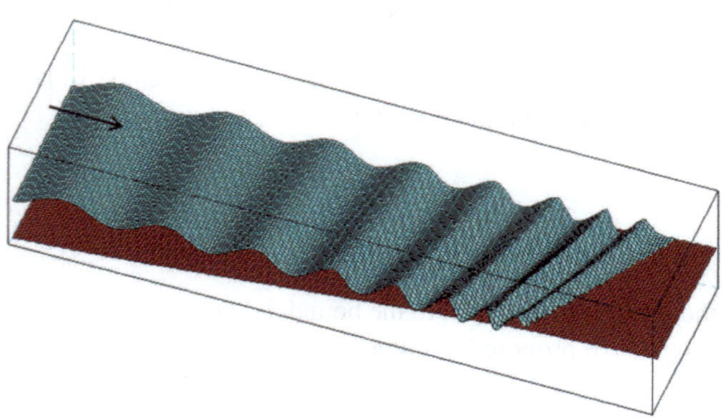

Fig. 5.6 Exercise 9. Snapshot of long waves being subject to refraction. The *blue surface* displays sea-level elevation. The *red surface* shows bathymetry

5.4 The Wind-Forced Shallow-Water Model

5.4.1 The Governing Equations

The shallow-water equations including wind-stress forcing and bottom friction read:

$$\begin{aligned}\frac{\partial u}{\partial t} &= -g\frac{\partial \eta}{\partial x} + \frac{\tau_x^{\text{wind}} - \tau_x^{\text{bot}}}{\rho_o h}\\ \frac{\partial v}{\partial t} &= -g\frac{\partial \eta}{\partial y} + \frac{\tau_y^{\text{wind}} - \tau_y^{\text{bot}}}{\rho_o h}\\ \frac{\partial \eta}{\partial t} &= -\frac{\partial (u\,h)}{\partial x} - \frac{\partial (v\,h)}{\partial y}\end{aligned} \tag{5.6}$$

where $(\tau_x^{\text{wind}}, \tau_y^{\text{wind}})$ is the wind-stress vector, and $(\tau_x^{\text{bot}}, \tau_y^{\text{bot}})$ is the frictional bottom-stress vector. For simplicity, lateral friction, the nonlinear terms, and the Coriolis force are not included yet in the momentum equations. This model only describes depth-averaged effects of wind forcing and bottom friction.

5.4.2 Semi-implicit Approach for Bottom Friction

Under the exclusive action of bottom friction and using a quadratic bottom-friction law, the momentum equation can be written as:

$$\frac{\partial u}{\partial t} = -ru\sqrt{(u^2+v^2)}/h \tag{5.7}$$

$$\frac{\partial v}{\partial t} = -rv\sqrt{(u^2+v^2)}/h \tag{5.8}$$

where r is a non-dimensional bottom-drag coefficient. Under the assumption that the initial flow runs into the x-direction, an explicit approach of the bottom-friction term would lead to the finite-difference equation:

$$u_{j,k}^{n+1} = u_{j,k}^{n}\,(1-\epsilon) \quad \text{with} \quad \epsilon = r\,\Delta t\,\left|u_{j,k}^{n}\right|/h_u$$

where h_u is thickness of the water column at the u-grid point. The problem now are instances of $\epsilon > 1$, which can happen in shallow parts of a model domain, triggering unwanted acceleration of the flow. Bottom friction cannot do such things. This problem can be avoided when using a semi-implicit approach for bottom friction, leading to the equations:

$$u_{j,k}^{n+1} = u_{j,k}^{n} - r\,\Delta t\; u_{j,k}^{n+1}\sqrt{\left(u_{j,k}^{n}\right)^2 + \left(v_u^n\right)^2}/h_u \tag{5.9}$$

$$v_{j,k}^{n+1} = v_{j,k}^{n} - r\Delta t\ v_{j,k}^{n+1}\sqrt{\left(u_{v}^{n}\right)^{2} + \left(v_{j,k}^{n}\right)^{2}} / h_{v} \tag{5.10}$$

where the subscripts u and v indicate the location at which a variable is calculated. This is necessary because u, v and h are not evaluated at the same grid point in the Arakawa C-grid (see Fig. 5.1). Reorganisation of these equations gives:

$$u_{j,k}^{n+1} = u_{j,k}^{n} / (1 + R_x) \tag{5.11}$$

$$v_{j,k}^{n+1} = v_{j,k}^{n} / \left(1 + R_y\right) \tag{5.12}$$

The parameters R_x and R_y are given by:

$$R_x = r\Delta t\sqrt{\left(u_{j,k}^{n}\right)^{2} + \left(v_{u}^{n}\right)^{2}} / h_u \tag{5.13}$$

$$R_y = r\Delta t\sqrt{\left(u_{v}^{n}\right)^{2} + \left(v_{j,k}^{n}\right)^{2}} / h_v \tag{5.14}$$

are always positive quantities, so that bottom friction will gradually decrease speed, as required. Hence, a semi-implicit approach for bottom friction should always be employed in layer models.

5.4.3 Finite-Difference Equations

Using a semi-implicit approach for bottom friction, the finite-difference equations stating momentum conservation are given by:

$$u_{j,k}^{n+1} = \left(u_{j,k}^{n} + \Delta u_{j,k}^{n}\right) / (1 + R_x) \tag{5.15}$$

$$v_{j,k}^{n+1} = \left(v_{j,k}^{n} + \Delta v_{j,k}^{n}\right) / \left(1 + R_y\right) \tag{5.16}$$

where R_x and R_y are given by (5.13) and (5.14), and

$$\Delta u_{j,k}^{n} = \Delta t\left\{\tau_x^{wind} / (\rho_o h_u) - g\left(\eta_{j,k+1}^{n} - \eta_{j,k}^{n}\right) / \Delta x\right\} \tag{5.17}$$

$$\Delta v_{j,k}^{n} = \Delta t\left\{\tau_y^{wind} / (\rho_o h_v) - g\left(\eta_{j+1,k}^{n} - \eta_{j,k}^{n}\right) / \Delta y\right\} \tag{5.18}$$

5.5 Exercise 10: Wind-Driven Flow in a Lake

5.5.1 Aim

The aim of this exercise is to simulate the wind-driven water circulation in a shallow lake with variable bottom topography.

5.5.2 Creation of Variable Bathymetry

Before simulating the wind-driven circulation in a shallow lake, I want to show the reader a simple way to create variable bottom topography without using complex analytical functions. The trick is to start with a coarse block-type bathymetry and to employ the diffusion equation for subsequent smoothing. The diffusion equation is given by:

$$\frac{\partial h}{\partial t} = A_h \left(\frac{\partial^2 h}{\partial x^2} + \frac{\partial^2 h}{\partial y^2} \right) \tag{5.19}$$

where the diffusion coefficient A_h and the duration of smoothing are adjusted such that the result is acceptable. Coastlines and land should not disappear during the process. This can be implemented in the code via the choice of zero-gradient conditions at the borders between dry and wet grid cells.

5.5.3 Sample Code

The FORTRAN 95 bathymetry creator, named "BathCreator.f95" is included in the folder "Miscellaneous/2D Bathymetry Creator" of the CD-ROM. The result is written to a file named "topo.dat". This file is required as input file for the simulation code. Included is also a SciLab script, called "Bath.sce", creating Figure 5.7.

5.5.4 Task Description

Consider a lake of 5 km×5 km in horizontal extent, being resolved by equidistant horizontal grid spacings of $\Delta x = \Delta y = 100$ m, and variable bathymetry. Figure 5.7 shows the lake's bathymetry used by the author. Consider a uniform southerly wind stress of $\tau_y^{\mathrm{wind}} = 0.2$ Pa in strength being linearly adjusted from zero to its final value over 2 days to avoid unwanted inertial and gravity waves. Simulate the steady-state circulation resulting from the wind-stress forcing applied. Five days of simulation should be sufficient for this. Set the bottom-friction coefficient to $r = 0.001$. I used a time step of $\Delta t = 3\,\mathrm{s}$, which satisfies the CFL stability criterion.

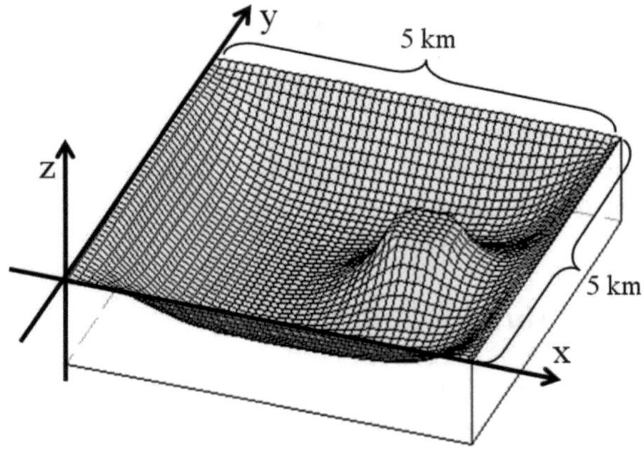

Fig. 5.7 Bathymetry for Exercise 10

5.5.5 *Tricks for Long Model Simulations*

This lake simulation might take a bit longer to complete. SciLab can be used to inspect preliminary results while the model is running in the background. If these results are unsatisfactory, simulations can be cancelled at any time by entering "<Ctrl> c" in the Command Prompt window.

5.5.6 *Results*

Flow speeds tend to be much swifter in shallow than in deeper regions of the domain: For graphical display purposes, circulations should be visualised by means of vertically integrated velocity components; that is, lateral volume transport. This removes the strong effect of water depth on flow speed and reveals the circulation pattern in deeper parts of the domain.

A steady-state circulation is established in the lake after 4 days of simulation (Fig. 5.8). Although the wind forcing is uniform, the circulation is quite complex. Owing to confinement by the lake's banks, not all of the flow can go into the wind direction. Instead of this, a return flow establishes moving water from the northwestern corner to the southeastern corner against the wind direction. This return flow is driven by (southeastward) pressure gradients associated with a slanting sea level. The island operates as a divider of the flow and strong lateral flow shear is found on the western side of the island. A counterclockwise "eddy" establishes northwest of the island. Maximum current speed is 0.7 m/s.

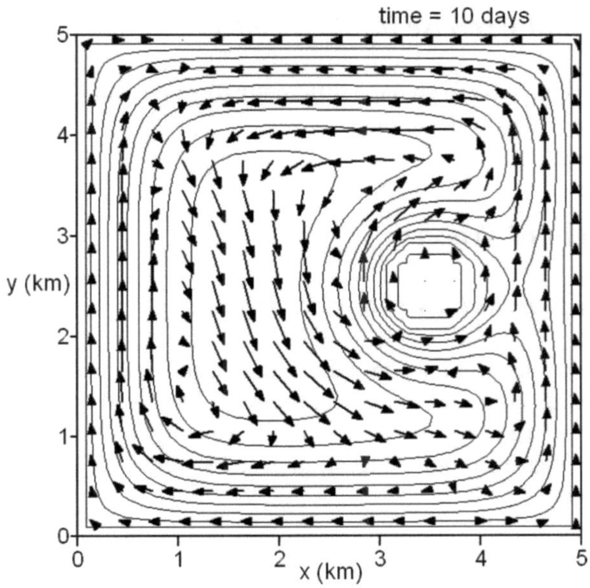

Fig. 5.8 Exercise 10. Steady-state currents in a shallow lake resulting from a southerly wind of a wind stress of 0.2 Pa. *Black lines* show bathymetric contours. *Arrows* are volume transport vectors ($u\ h$,$v\ h$) being averaged over 3×3 grid cells

5.5.7 Sample Code and Animation Script

The folder "Exercise 10" of the CD-ROM contains the computer codes. The file "info.txt" gives additional information. Note that the output flow field is needed as input data for Exercises 11 and 12.

5.5.8 Caution

The reader will notice that the simulation takes several minutes to complete. With the choice of finer grid spacings, the simulation time will significantly increase. Using half the grid spacing, for instance, has two consequences. First, the number of grid cells increases fourfold and so does the time to complete each simulation loop. Second, smaller time steps are required to match the CFL stability criterion. Both features together, will lead to an eightfold increase in total simulation time. Also the frequency of data outputs needs to be carefully chosen. I used outputs at every second hour of iteration, yielding $12 \times 5 = 60$ outputs per variable. The resultant size of each data file is $>$1.8 MByte. Keep the frequency of data outputs as low as possible to avoid storage space problems.

5.5.9 Additional Exercise for the Reader

The reader is encouraged to create a different bathymetry and to explore the lake's circulation for a variety of wind directions.

5.6 Movement of Tracers

5.6.1 Lagrangian Versus Eulerian Tracers

Lagrangian tracers are non-buoyant fluid parcels that move passively with the flow. In practice, the trajectory of Lagrangian floats are predicted by means of displacement distances per time step derived from the velocity at float locations. Eulerian tracers, on the other hand, are concentration fields being subject to advection and mixing by currents.

5.6.2 A Difficult Task

The numerical simulation of advection of Eulerian concentration fields is a challenging and difficult task and there are many potential sources of errors that can occur. Some numerical advection schemes trigger unwanted numerical diffusion, while other schemes lead to unwanted numerical oscillations.

5.6.3 Eulerian Advection Schemes

The advection equation for depth-averaged tracer concentration B in the presence of depth-averaged horizontal flow with components u and v is given by:

$$\frac{\partial B}{\partial t} = -u\frac{\partial B}{\partial x} - v\frac{\partial B}{\partial y} \tag{5.20}$$

Using the product rule of differentiation, this equation can be reformulated as:

$$\frac{\partial B}{\partial t} = -\frac{\partial (uB)}{\partial x} - \frac{\partial (vB)}{\partial y} + B\left(\frac{\partial u}{\partial x} + \frac{\partial v}{\partial y}\right) \tag{5.21}$$

The temporal change of B can be discretised using a simple time-forward iteration. Also the last term can be formulated in a straight-forward explicit manner. Several options are available for the treatment of the remaining terms. These options are described in the following for the x-direction. Analog recipes apply for the y-direction.

In all schemes, the first term on the right-hand side of the latter equation can be formulated as:

$$-\Delta t \frac{\partial(uB)}{\partial x} = C_w B_w - C_e B_e \tag{5.22}$$

where the indices "w" and "e" refer to east and west faces of the control volume and:

$$C_w = u_{k-1}^n \Delta t/\Delta x \;\text{ and }\; C_e = u_k^n \Delta t/\Delta x \tag{5.23}$$

are so-called *Courant numbers*. In a next step, we can split the u component into positive and negative components:

$$u^+ = 0.5(u + |u|) \;\text{ and }\; u^- = 0.5(u - |u|) \tag{5.24}$$

and rewrite (5.21) in the form:

$$-\Delta t \frac{\partial(uB)}{\partial x} = C_w^+ B_w^+ + C_w^- B_w^- - C_e^+ B_e^+ - C_e^- B_e^- \tag{5.25}$$

The objective of any finite-difference Eulerian advection scheme is to interpolate the volume-averaged values of B to obtain the effective face values B_e and B_w. Here we use so-called *Total Variation Diminishing* schemes or TVD schemes that are based on the requirement:

$$\sum_k \left| B_{k+1}^{n+1} - B_k^{n+1} \right| \le \sum_k \left| B_{k+1}^n - B_k^n \right| \tag{5.26}$$

For the TVD schemes used here, described by Fringer *et al.* (2005), the face values of B are computed with the upwind values plus the addition of a higher order term with:

$$\begin{aligned}
B_e^+ &= B_k^n + 0.5\Psi\left(r_k^+\right)\left(1 - C_e^+\right)\left(B_{k+1}^n - B_k^n\right) \\
B_e^- &= B_{k+1}^n - 0.5\Psi\left(r_k^-\right)\left(1 + C_e^-\right)\left(B_{k+1}^n - B_k^n\right) \\
B_w^+ &= B_{k-1}^n + 0.5\Psi\left(r_{k-1}^+\right)\left(1 - C_w^+\right)\left(B_k^n - B_{k-1}^n\right) \\
B_w^- &= B_k^n - 0.5\Psi\left(r_{k-1}^-\right)\left(1 + C_w^-\right)\left(B_k^n - B_{k-1}^n\right)
\end{aligned}$$

where the r parameters are given by:

$$r_k^+ = \frac{B_k^n - B_{k-1}^n}{B_{k+1}^n - B_k^n} \;\text{ and }\; r_k^- = \frac{B_{k+2}^n - B_{k+1}^n}{B_{k+1}^n - B_k^n}$$

The limiting function Ψ defines the particular scheme that is used. A few selected options are given in the following.

- The upstream scheme follows from $\Psi = 0$. We have used this scheme in the prediction of sea-level elevation in previous exercises.
- $\Psi = 1$ gives the Lax-Wendroff scheme.
- $\Psi(r) = \max\{0, \min(2r, 1), \min(r, 2)\}$ defines the so-called Superbee scheme.

In addition to this, we consider the Super-C scheme that is boundless by using the Courant number in the denominator. This scheme is defined by:

$$\Psi(r, |C|) = \begin{cases} \min(2r/|C|, 1) & : \quad 0 \le r \le 1 \\ \min(r/2/(1 - |C|), r) & : \quad r > 1 \\ 0 & : \quad \text{otherwise} \end{cases}$$

where the Courant number C is calculated at the right-hand face of a control volume.

5.6.4 *Stability Criterion for the Advection Equation*

The stability criterion for the above explicit forms of the advection equation is:

$$C = \frac{\Delta t}{\Delta x} u \le 1 \tag{5.27}$$

where C is the Courant number, and u is the flow speed. Accordingly, time steps have to satisfy the condition:

$$\Delta t \le \frac{\Delta x}{u} \tag{5.28}$$

Note that this is also a CFL condition, but this time based on flow speed instead of phase speed of waves. Conditions of $C < 1$ will always lead to a certain level of numerical diffusion. This is because the displacement distance per time step is less the grid spacing, so that some averaging will take place. Nevertheless, a particular advection scheme might perform better than others, which will be investigated in the following.

5.7 Exercise 11: Eulerian Advection

5.7.1 *Aim*

The aim of this exercise is to simulate the movement of non-buoyant Eulerian tracer subject to the steady-state lake's circulation predicted in Exercise 10. Different TVD advection schemes will be tested.

5.7.2 Task Description

We use the steady-state flow field computed in Exercise 10 to predict the movement pattern of Eulerian tracer being introduced at a concentration of unity in a certain region of the model domain. In this exercise, tracer is released in the northwestern part of the lake in a quadratic box with side lengths of 0.5 km. With an inspection of the steady-state circulation established in the lake (see Fig. 5.8), we expect that this tracer is initially advected southward and separates into westward and eastward flowing branches near the southern boundary. With "frozen" dynamics, a time step much greater compared with that in Exercise 10 can be used. I used $\Delta t = 200$ s, which satisfies the CFL condition (5.4). Advection schemes being tested are the upstream scheme, the Lax-Wendroff scheme, the Superbee scheme and the Super-C scheme.

5.7.3 Results

Figure 5.9 shows results employing the upstream scheme. This scheme is extremely numerically diffusive and triggers substantial artificial decrease of the maximum concentration by 85% after 9 h of simulation corresponding to 162 simulation steps. Owing to this numerical diffusion, tracer concentration is vigorously mixed horizontally.

The Lax-Wendroff scheme appears to be less diffusive (Fig. 5.10), but has other disadvantages. This scheme produces numerical oscillations, leading to slightly negative concentrations in some regions and concentrations exceeding the initial concentration in other regions.

The Superbee scheme produces more convincing results (Fig. 5.11) void of numerical oscillations and far less diffusive compared with the upstream scheme.

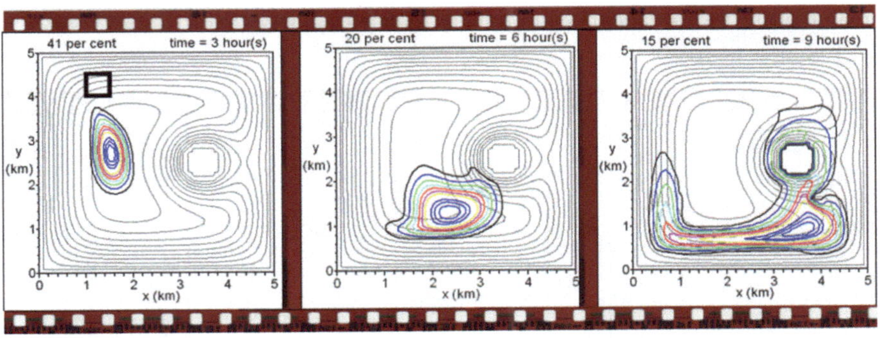

Fig. 5.9 Exercise 11. Snapshots of contours of tracer concentration (coloured lines) using the upstream scheme. The contour interval is maximum tracer concentration divided by 10. The header displays maximum concentration relative to initial concentration in per cent. *Thin black lines* are bathymetric contours. The *square* indicates the release area of the tracer

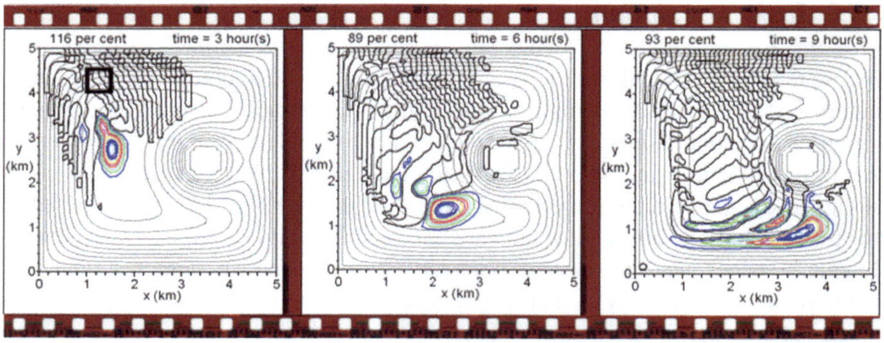

Fig. 5.10 Exercise 11. Same as Fig. 5.9, but with use of the Lax-Wendroff scheme

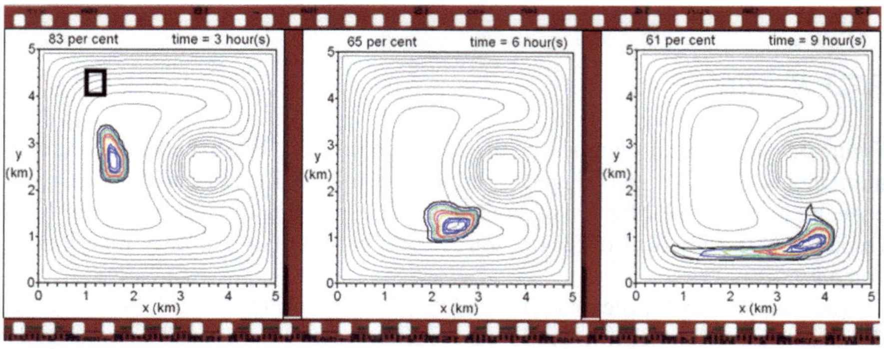

Fig. 5.11 Exercise 11. Same as Fig. 5.9, but with use of the Superbee limiter

After 9 h of simulation, the initial concentration has reduced by 39%, which is half the diffusion rate produced by the upstream scheme.

The Super-C scheme produces results similar to the Superbee scheme (Fig. 5.12). It induces only little diffusion and is largely void of numerical oscillations. After 9 h of simulation, the initial concentration has reduced by only 19%.

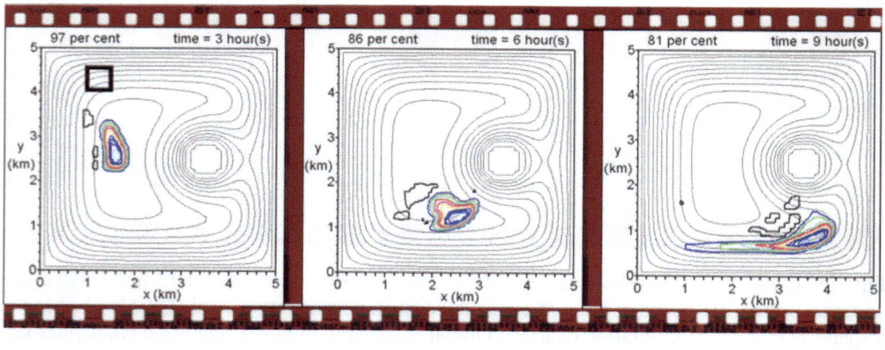

Fig. 5.12 Exercise 11. Same as Fig. 5.9, but with use of the Super-C scheme

5.7.4 Recommendation

Use either the Superbee limiter or the Super-C scheme for advection of Eulerian tracer. Either of these schemes should also be used in replacement of the upstream scheme in the volume-conservation equation.

5.7.5 Sample Code and Animation Script

The folder "Exercise 11" of the CD-ROM contains the computer codes for this exercise. The "MODE" switch allows for selection of any of the above flux limiters.

5.8 Exercise 12: Trajectories

5.8.1 Aim

The aim of this exercise is to predict the pathways of individual non-buoyant Lagrangian floats subject to the lake's steady-state circulation circulation.

5.8.2 Task Description

Using the steady-state flow field predicted in Exercise 10, a large number (3000) of Lagrangian floats is introduced at random locations in the lake to predict their pathways over a day. The horizontal displacement of a float is calculated from:

$$X_m^{n+1} = X_m^n + \Delta t \ U_m^n$$
$$Y_m^{n+1} = Y_m^n + \Delta t \ V_m^n$$

where m is the float number, X and Y specifies the location of a float, and U and V is the ambient lateral flow interpolated to the float location. To make this task easier, instead of interpolating velocity to the precise location of a float, we use the velocity interpolated to the nearest "h" grid point as a proxy. Zones within a distance of 500 m from the lake's banks are initially kept free of floats to avoid that floats become trapped in zones of little or zero flow.

5.8.3 Results

Both the animation movie of float locations (Fig. 5.13 shows a snapshot) and trajectories (Fig. 5.14) nicely reveal the circulation pattern established in the lake. The southerly wind drives northward flows on the western and eastern sides of the lake. In interaction with bathymetry, the resultant pattern in sea-level gradients creates a

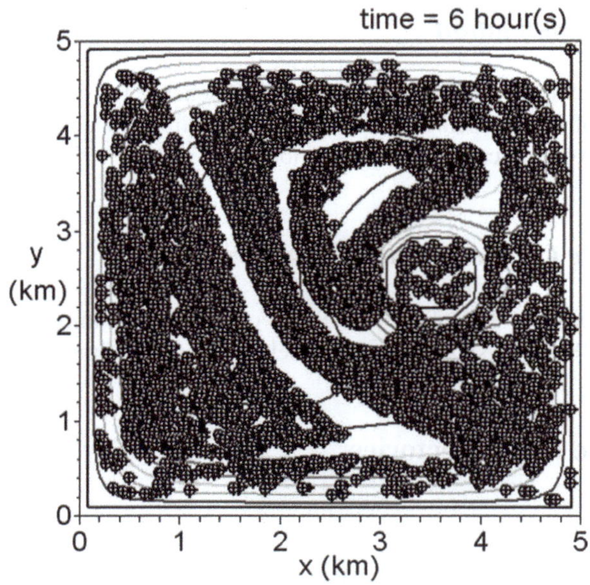

Fig. 5.13 Exercise 12. Snapshot of the locations of 3000 Lagrangian floats after 6 h of simulation

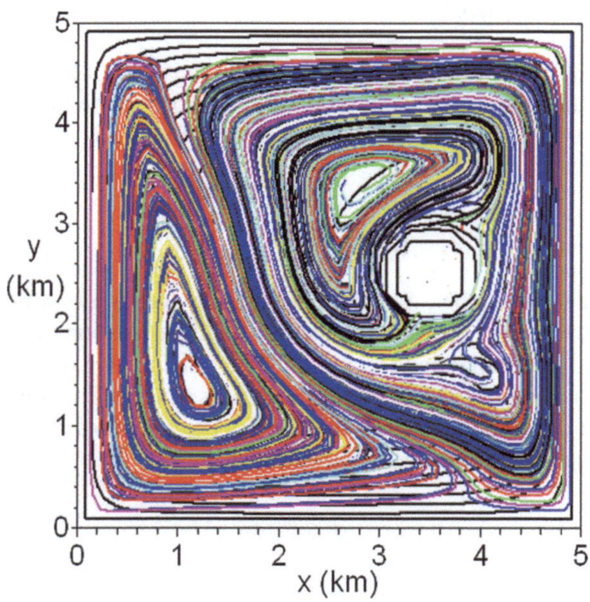

Fig. 5.14 Exercise 12. Trajectories of 500 Lagrangian floats over 12 h of simulation

return flow running from the northwestern corner to the southeastern corner of the lake. This return flow disintegrates into two separate "gyres" as it approaches the southern bank of the lake. Obviously, the island forms an obstacle for the eastern gyre.

5.8.4 Sample Code and Animation Script

The computer codes for this exercise can be found in the folder "Exercise 12" of the CD-ROM. The code includes a random-number generator, taken from Press *et al.* (1989), for allocation of initial float locations. One Scilab script produces an animation of the drift of floats, whereas the other script produces a single graph displaying trajectories of a selected number of floats (see Fig. 5.14).

5.9 Exercise 13: Inclusion of Nonlinear Terms

5.9.1 Aim

The aim of this exercise is to include the nonlinear terms (advection of momentum) in the shallow-water equations.

5.9.2 Formulation of the Nonlinear Terms

Using the product rule of differentiation, the (horizontal) nonlinear terms in our shallow-water model can be written as:

$$\mathrm{Adv}_h(\xi) = u\frac{\partial \xi}{\partial x} + v\frac{\partial \xi}{\partial y} = \frac{\partial (u\xi)}{\partial x} + \frac{\partial (v\xi)}{\partial y} - \xi\left(\frac{\partial u}{\partial x} + \frac{\partial v}{\partial y}\right) \tag{5.29}$$

where ξ is either u or v. The first two terms on the right-hand side of this equation can be discretised using the TVD advection schemes for a control volume as in Exercise 11. The remaining term can be formulated in an explicit manner.

5.9.3 Sample Code

Due to its multiple use, it make sense to formulate the advection scheme in generalised form as a subroutine. This subroutine can then be used for calculations of the nonlinear terms and advection of Eulerian tracer, and also as a solver of the vertically integrated form of the continuity equation. In this exercise, the Superbee scheme is used for the sea-level predictor and different flux limiters are tested for the nonlinear terms. The folder "Exercise 13" of the CD-ROM contains the amended simulation code.

5.9.4 Results

Inclusion of the nonlinear terms modifies the lake's circulation (Fig. 5.15). A clockwise eddy establishes in the northwestern corner of the lake and the eddy northwest of the island has largely disappeared. The upstream scheme is highly diffusive and therefore triggers rapid establishment of a steady-state circulation in the lake. In contrast to this, less diffusive TVD advection schemes based on either the Superbee or the Super-C limiters lead to slight oscillations of the circulation pattern presumably triggered by the initial adjustment of the wind field. Due to reduced numerical diffusion, either of these schemes should be applied for the nonlinear terms. Use of the Lax-Wendroff scheme for the nonlinear terms did not lead to satisfactory results.

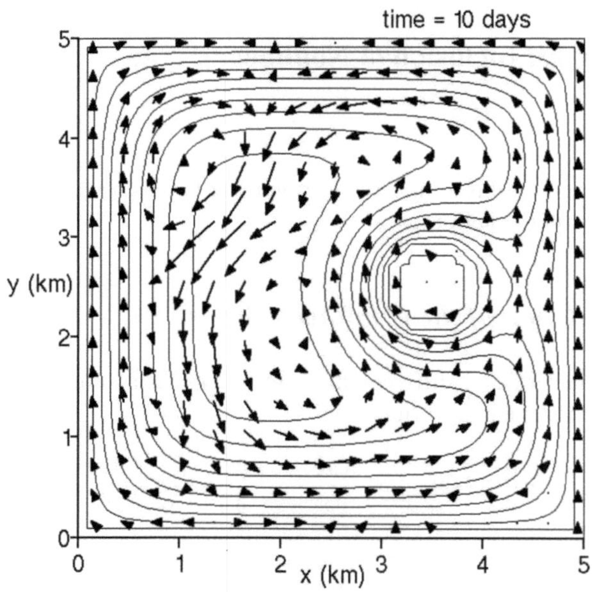

Fig. 5.15 Exercise 13. Same as Fig. 5.8, but with inclusion of nonlinear terms in the momentum equations using the TVD Superbee scheme

5.10 Exercise 14: Island Wakes

5.10.1 Aim

The aim of this exercise is to simulate turbulent wakes produced by horizontal flows around an island. This includes implementations of both lateral friction and lateral momentum diffusion in the shallow-water model.

5.10.2 The Reynolds Number

Flow around an obstacle such as an island becomes dynamically unstable under certain circumstances and breaks up into a irregular turbulent wake. The transition of laminar flow into turbulence can be described by means of the ratio between the nonlinear terms and diffusion of momentum. This ratio is called the *Reynolds number* (Reynolds, 1883) and can be defined by:

$$Re = \frac{UL}{A_h} \tag{5.30}$$

where U is the speed of the incident flow, L is the diameter of the obstacle, and A_h is ambient horizontal eddy viscosity.

A variety of flow regimes can develop in dependence on the magnitude of the Reynolds number. For $Re \approx 1$ the flow is typically laminar and smoothly surrounds the obstacle. A stationary vortex pair with central return flow develops for $Re \approx 10$. Larger values of $Re \approx 100$ leads to the formation of a turbulent wake in the lee of the obstacle. $Re >> 100$ triggers a turbulent wake of organised vortices called *von Kármán vortex* shedding in appreciation of pioneering work by Theodore von Kármán (1911).

Vortex shedding occurs at a certain frequency f. The dimensioness number:

$$St = \frac{fL}{U} \tag{5.31}$$

is known as the Strouhal number and is named after the Czech physicist Vincenc Strouhal (1850–1922), who first investigated the steady humming (or singing) of telegraph wiring. There exist relationships (not replicated here) between the Strouhal number and the Reynolds number that can be experimentally derived. It should be noted that this Strouhal instability is believed to be the reason for collapse of the Tacoma Narrows Bridge, Washington, on November 7, 1940.

5.10.3 Inclusion of Lateral Friction and Momentum Diffusion

Lateral friction and diffusion of momentum is required in the momentum equations in order to simulate the development of turbulent wakes in the lee of an obstacle. Under the assumption of uniform values of lateral eddy viscosity A_h, the depth-averaged version of the lateral momentum diffusion can be formulated as:

$$\mathrm{div}_h(u) = \frac{A_h}{h}\left\{\frac{\partial}{\partial x}\left(h\frac{\partial u}{\partial x}\right) + \frac{\partial}{\partial y}\left(h\frac{\partial u}{\partial y}\right)\right\} \tag{5.32}$$

$$\mathrm{div}_h(v) = \frac{A_h}{h}\left\{\frac{\partial}{\partial x}\left(h\frac{\partial v}{\partial x}\right) + \frac{\partial}{\partial y}\left(h\frac{\partial v}{\partial y}\right)\right\} \tag{5.33}$$

where h is layer thickness. The discretised form of the diffusion term for u has the components:

$$\frac{A_h}{h}\frac{\partial}{\partial x}\left(h\frac{\partial u}{\partial x}\right) = \frac{A_h}{h_u \Delta x}\left[\frac{h_e\left(u^n_{j,k+1} - u^n_{j,k}\right)}{\Delta x} - \frac{h_w\left(u^n_{j,k} - u^n_{j,k-1}\right)}{\Delta x}\right] \tag{5.34}$$

$$\frac{A_h}{h}\frac{\partial}{\partial y}\left(h\frac{\partial u}{\partial y}\right) = \frac{A_h}{h_u \Delta y}\left[\frac{h_n\left(u^n_{j+1,k} - u^n_{j,k}\right)}{\Delta y} - \frac{h_s\left(u^n_{j,k} - u^n_{j-1,k}\right)}{\Delta y}\right] \tag{5.35}$$

where layer thicknesses of the faces of the control volume (see Fig. 5.16) are given by:

$$\begin{aligned}
h_u &= 0.5\left(h^n_{j,k} + h^n_{j,k+1}\right)\\
h_e &= h^n_{j,k+1}\\
h_w &= h^n_{j,k}\\
h_n &= 0.25\left(h^n_{j,k} + h^n_{j,k+1} + h^n_{j+1,k} + h^n_{j+1,k+1}\right)\\
h_s &= 0.25\left(h^n_{j,k} + h^n_{j,k+1} + h^n_{j-1,k} + h^n_{j-1,k+1}\right)
\end{aligned}$$

The diffusion terms for v are constructed in a similar fashion.

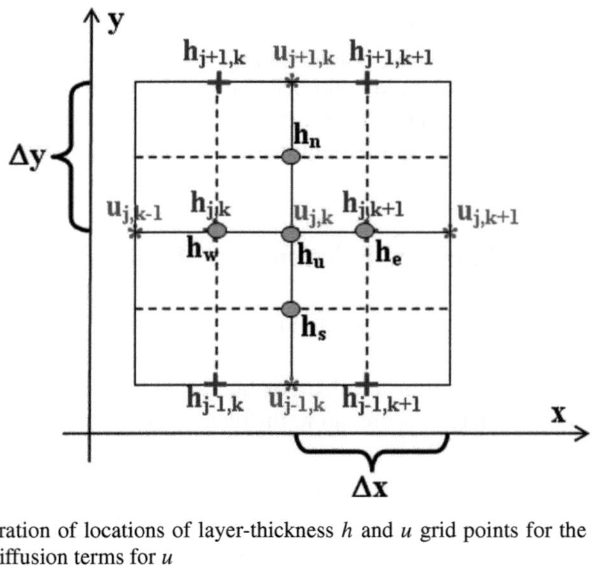

Fig. 5.16 Illustration of locations of layer-thickness h and u grid points for the discretisation of the horizontal diffusion terms for u

5.10.4 Stability Criterion for Diffusion Terms

The one-dimensional diffusion equation for a variable ψ can be written as:

$$\frac{\partial \psi}{\partial t} = A_h \frac{\partial^2 \psi}{\partial x^2} \tag{5.36}$$

where A_h is a diffusivity assumed constant. Using an explicit finite-difference formulation of diffusion term leads to the stability criterion:

$$\Delta t \leq \frac{(\Delta x)^2}{A_h} \tag{5.37}$$

Although the diffusion terms (5.31) and (5.31) in the momentum equations are slightly more complex than assumed here, the latter condition gives a useful upper bound for permitted time steps. If problems persist, the time step should be further reduced.

5.10.5 Full-Slip, Semi-Slip and No-Slip Conditions

Frictional effects on flow running parallel to coastlines can be implemented via specification of the velocity shear near the coast. For, instance, with the choice of zero-gradient conditions for this velocity component, there is no shear of flow parallel to the coast and, accordingly, there is no frictional stress imposed on the flow. This is called the *full-slip condition.*

As another option, flow vanishes at the coastline under the assumption that the flow at the grid point on the other side of the coastline is anti-parallel to the coastal flow, so that the average value vanishes directly at the coast. This is known as the

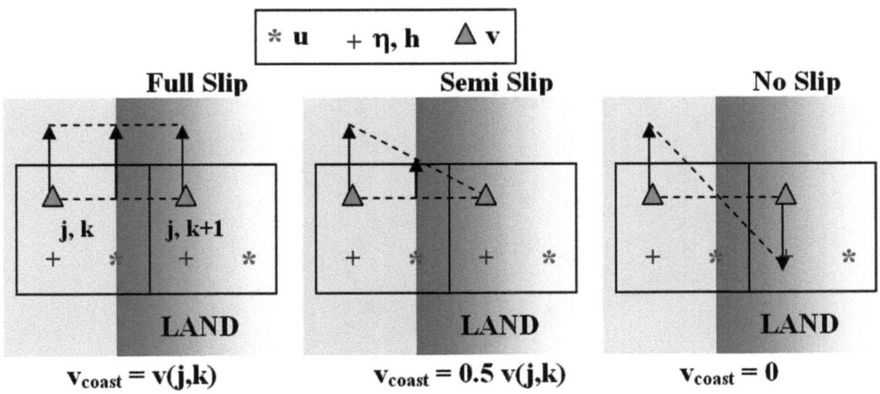

Fig. 5.17 Illustration of the full-slip, semi-slip and no-slip conditions used for flow parallel to coastlines

zero-slip condition. The *semi-slip condition* gives half the velocity shear compared with the no-slip condition and it is realised by setting flow at land grid points to zero. Figure 5.17 illustrates these conditions.

5.10.6 Task Description

Employ the bathymetry creator for construction of a bathymetry similar to that in Fig. 5.18. The western and eastern boundaries of the model domain are open and a small island is located near the western boundary. Use nx = 101 grid cells in the x-direction and ny = 51 grid points in the y direction together with equidistant grid spacings of $\Delta x = \Delta y = 100$ m. Use a numerical time step of $\Delta t = 3$ s.

Add the horizontal diffusion terms to the momentum equations and choose a no-slip condition. The model is forced by prescription of a westerly (eastward) wind stress of $\tau_x^{\text{wind}} = 0.2$ Pa. The TVD Superbee scheme is used for advection of any property. All other parameters are the same as in Exercise 13. In this exercise, the eastern and western boundaries are cyclic boundaries. This means that fluid escaping through the eastern boundary enters the western boundary and vice versa.

In anticipation of eastward flow, a point source of Eulerian tracer concentration is introduced at the western boundary for visualisation of the flow dynamics. In contrast to the dynamical variables, Eulerian tracer is allowed to disappear through the eastern boundary using zero-gradient conditions at this boundary.

The wind-forcing imposed will create an incident flow of about 0.5 m/s in speed. The diameter of the island is 300 m. Consider the following cases. Case 1 uses $A_h = 2.5$ m^2/s, giving $Re = 60$. Case 2 uses $A_h = 1$ m^2/s, yielding $Re = 300$. Run these two cases over 2 days with data outputs at hourly interval.

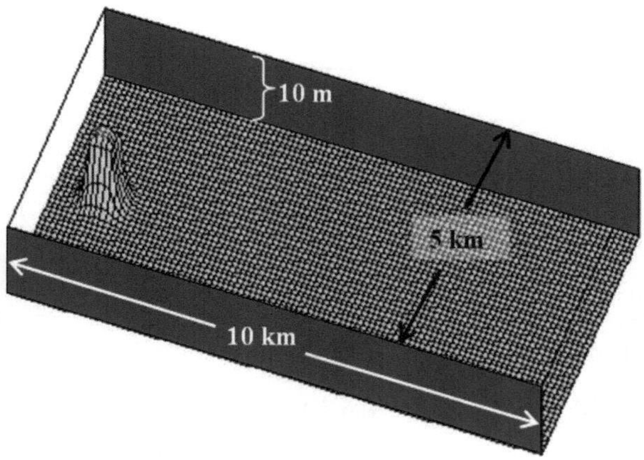

Fig. 5.18 Bathymetry for Exercise 14

5.10.7 Sample Code

The folder "Exercise 14" contains the computer code including implementations of lateral momentum diffusion and lateral friction with coastlines. Different slip conditions can be selected by means of a switch called "slip".

5.10.8 Results

In case of a relatively low Reynolds number of $Re = 60$ (Case 1), the flow around the island is relatively smooth (Fig. 5.19). A return flow is created behind the island with a speed of 8 cm/s. Note that the averaging procedure used in the SciLab script has removed this return flow in Fig. 5.19. The flow develops some slight meandering in the lee of the island toward the end of the simulation, but a turbulent wake is not created. It should be noted that numerical diffusion, being difficult to quantify, also contributes to lateral diffusion of momentum. This can hinder the formation of a turbulent wake.

A turbulent wake in form of a von Kármán vortex street forms for a higher Reynolds number of $Re = 300$ (Case 2) (Fig. 5.20). Vortices attain a lengthscale of the order of the island's diameter and produce lateral mixing in the lee of the island. This model application captures well the transition of largely laminar flow for small Reynolds numbers into a vortex street for $Re > 100$.

The generation mechanism of this process can be described as follows. The flow is forced around the island by pushing water against the island. This produces elevated sea level and associated pressure gradients divert the flow around the island. Similarly, movement of water away from the island in its lee leads to a local drop of the sea level. For weak flows, this low-pressure centre forces the flow smoothly around the island. For stronger flows, this low pressure centre intensifies

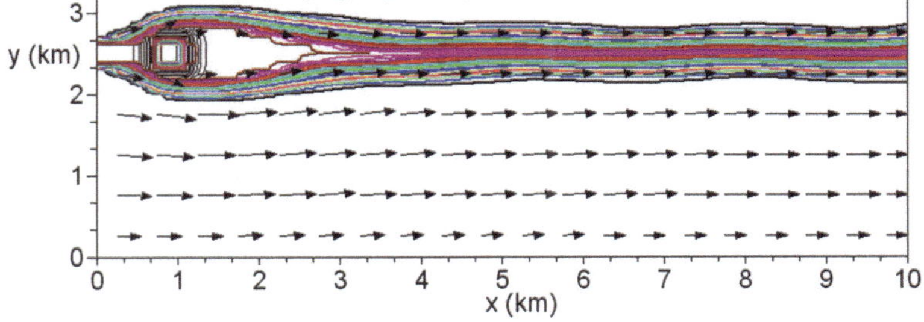

Fig. 5.19 Exercise 14. Case 1 ($Re = 60$). Flow vectors (averaged over 5×5 grid cells) and Eulerian tracer concentration (lines, contour interval is 0.04) after 2 days of iteration

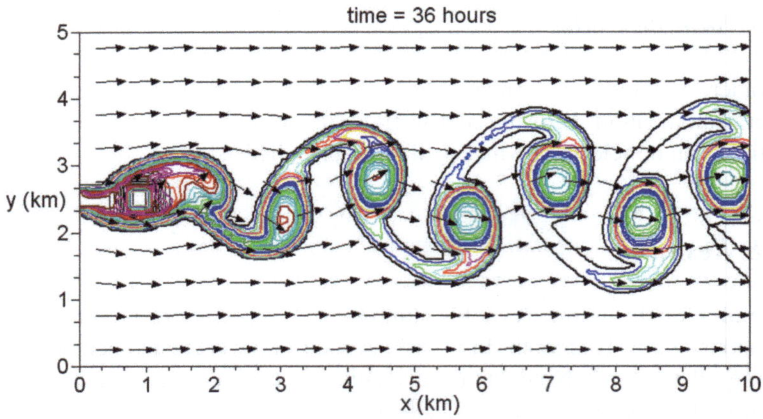

Fig. 5.20 Exercise 14. Same as Fig. 5.19, but for Case 2 ($Re = 300$) and after 1.5 days of simulation

and produces a return flow toward the island and an associated counter-rotating vortex pair. With a further intensification of the flow incident on the island, the vortex pair becomes unstable owing to nonlinear effects. This creates flow meadering and the formation of a vortex street. The role of lateral momentum diffusion in this process is the suppression of flow disturbances which fails when the ambient flow exceeds a certain threshold speed.

5.10.9 Additional Exercises for the Reader

Explore changes in the dynamics when using either the semi-slip or the full-slip condition. Experiment with different island shapes and sizes. Place two or more islands in the model domain to study interference patterns of adjacent turbulent wakes.

Chapter 6
Rotational Effects

Abstract This chapter applies the shallow-water equations to study a variety of hydrodynamic processes being influenced or even controlled by the Coriolis force. The reader is introduced to quasi-geostrophic flows and the concept of vorticity. Exercises address a rich variety of processes including coastal Kelvin waves, topographic steering of barotropic quasi-geostrophic flows, topographic Rossby waves, the general wind-driven circulation of the ocean, western boundary currents, baroclinic compensation, geostrophic adjustment of density fronts, the baroclinic instability mechanism, and reduced-gravity plumes.

6.1 The Complete Shallow-Water Equations

6.1.1 Description

Exercises in this chapter employ the single-layer shallow-water equations in their complete form including nonlinear terms, wind-stress forcing, the Coriolis force, the pressure-gradient force, bottom friction and lateral diffusion of momentum. The momentum equations take the form:

$$\frac{\partial u}{\partial t} + \mathrm{Adv}_h(u) - f v = -g\frac{\partial \eta}{\partial x} + \frac{\tau_x^{wind} - \tau_x^{bot}}{\rho_o h} + \mathrm{Diff}_h(u) \tag{6.1}$$

$$\frac{\partial v}{\partial t} + \mathrm{Adv}_h(v) + f u = -g\frac{\partial \eta}{\partial y} + \frac{\tau_y^{wind} - \tau_y^{bot}}{\rho_o h} + \mathrm{Diff}_h(v) \tag{6.2}$$

where Adv_h denotes the nonlinear terms, given by (5.29), and Diff_h the lateral friction terms, being of the form of (5.31). The Coriolis force appears as a new force in these equations.

6.1.2 Implementation of the Coriolis Force

The shallow-water equations are solved in the following steps.

J. Kämpf, *Ocean Modelling for Beginners*,
DOI 10.1007/978-3-642-00820-7_6, © Springer-Verlag Berlin Heidelberg 2009

Step 1: Predict a first-guess velocity $(u^*_{j,k}, v^*_{j,k})$ without the Coriolis force but a semi-implicit approach for bottom friction, as done in Exercise 10.

Step 2: Apply the semi-implicit approach for the Coriolis force, which leads to the equations:

$$u^{n+1}_{j,k} = \left[u^*_{j,k} - \beta u^n_{j,k} + \alpha v^n_{j,k}\right]/(1+\beta)$$
$$v^{n+1}_{j,k} = \left[v^*_{j,k} - \beta v^n_{j,k} - \alpha u^n_{j,k}\right]/(1+\beta)$$

where $\alpha = \Delta t f$ and $\beta = 0.25\,\alpha^2$. To be able to use the flooding algorithm of previous exercises, velocity changes are calculated from:

$$\Delta u_{j,k} = u^{n+1}_{j,k} - u^n_{j,k}$$
$$\Delta v_{j,k} = v^{n+1}_{j,k} - v^n_{j,k}$$

before update of the velocity field is updated. As in previous model codes, the predicted velocity components are used as input for the sea-level predictor.

6.2 Coastal Kelvin Waves

6.2.1 Theory

Coastal Kelvin waves are of the form of surface or interfacial gravity waves that under the influence of the Coriolis force travel along a coastline with maximum amplitudes at the coast. The description of such waves can be traced back to Sir William Thomson (later to become Lord Kelvin) (Thomson, 1879). The simplest way to analytically describe the dynamics of Kelvin waves is to consider a constant-density coastal ocean of constant depth H, bounded by a straight coastline aligned with the x-direction, and to request absence of any onshore or offshore flow. In this case, the linear, frictionless shallow-water equations take the form:

$$\frac{\partial u}{\partial t} = -g\frac{\partial \eta}{\partial x}$$
$$f u = -g\frac{\partial \eta}{\partial y} \tag{6.3}$$
$$\frac{\partial \eta}{\partial t} = -\frac{\partial (uH)}{\partial x}$$

where x is the alongshore direction and positive y denotes the offshore direction. The wave solutions of these equations for small-amplitude disturbances ($\eta << H$) are given by:

$$\eta(x, y, t) = \eta_o \exp(-y/R) \sin(kx - \omega t) \tag{6.4}$$

$$u(x, y, t) = \sqrt{\frac{g}{H}}\, \eta_o \exp(-y/R) \sin(kx - \omega t) \tag{6.5}$$

where $k = 2\pi/\lambda$ (λ is wavelength), $\omega = 2\pi/T$ (T is wave period), and R is the *Rossby radius of deformation*, defined by:

$$R = \frac{\sqrt{gH}}{|f|} \tag{6.6}$$

It can also be shown that these waves travel with a phase speed of long surface gravity waves of $c = \sqrt{gH}$ along the coast with the coast on their right in the northern hemisphere and on their left in the southern hemisphere. Their amplitude is maximum at the coast and decreases exponentially away from the coast on a lengthscale of the deformation radius.

6.3 Exercise 15: Coastal Kelvin Waves

6.3.1 Aim

The aim of this exercise is to simulate the structure and dynamics of coastal Kelvin waves.

6.3.2 Task Description

Consider a rectangular model domain with a length of 400 km, a width of 100 km, and a depth of 10 m. Lateral grid spacing is set to $\Delta x = \Delta y = 2$ km. The time step is set to $\Delta t = 10$ s. All boundaries are treated as coasts. An unrealistically high value of the Coriolis parameter of $f = +5 \times 10^{-4}\ \text{s}^{-1}$ is chosen to reduce the deformation radius to $R \approx 20$ km and to keep the total simulation time within a reasonable limit. The associated inertial period is 3.5 h. The smart reader will jump on the table and claim that the minimum inertial period on Earth is 12 hours. This is true, but with a model we can do a little bit of science fiction, can't we?

A wave paddle is located near the lower left corner of the model domain oscillating the sea level with an amplitude of 1 m and a period of 2 hours. This period is not far away from the fictional inertial period, so that an influence by the Coriolis force can be anticipated. The task is to simulate the resultant wave field over half a day (or 3.4 inertial periods) with outputs of sea-level elevation and velocity fields at every 10 mins.

The first-order Shapiro filter with a smoothing parameter of $\epsilon = 0.05$ should be applied. For simplicity, wind-stress forcing, bottom friction and lateral momentum diffusion are disabled. The TDV Superbee advection scheme is used for both the nonlinear terms and in the vertically integrated continuity equation.

6.3.3 Results

Figure 6.1 shows the propagation of a coastal Kelvin wave attaining maximal amplitude at the coast and showing an exponential decrease in amplitude away from the coast. Numerical diffusion is apparent resulting in a decrease of wave amplitude along the coast. Storm surges can stimulate such coastal Kelvin waves in shelf seas of horizontal dimensions exceeding the Rossby radius of deformation. The North Sea is one example of this.

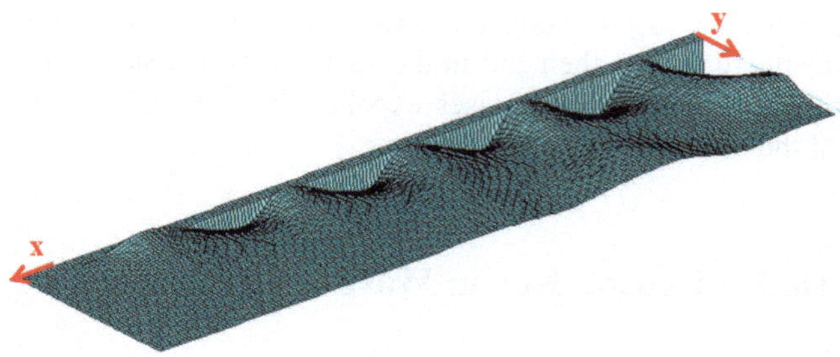

Fig. 6.1 Exercise 15. Snapshot of the deformation of the sea surface caused by a coastal Kelvin wave in the northern hemisphere

6.3.4 Sample Codes and Animation Script

The folder "Exercise 15" on the CD-ROM contains the computer codes for this exercise.

6.3.5 Additional Exercise for the Reader

Repeat this exercise with different values of the Coriolis parameter and total water depth and explore the resultant wave patterns. The reader might also introduce some bottom friction.

6.4 Geostrophic Flow

6.4.1 Scaling

If we use the Coriolis force for reference, we can define various force ratios that essentially compare different time scales. The temporal Rossby number (see

Sect. 3.17) compares the inertial period with the time scale of a process. The Coriolis force influences or even controls processes that have time scales of or exceeding the inertial period. On the other hand, the ratio between the nonlinear terms and the Coriolis force is called the *Rossby number* and is defined by:

$$Ro = \frac{U}{fL} \tag{6.7}$$

where U is a typical speed, f is the inertial period, and L is the lengthscale of a process. Again, this is a comparison of time scales, whereby L/U is the time it takes for a flow of speed U to travel a distance of L. For small Rossby numbers ($Ro << 1$), nonlinear terms are negligibly small compared with the Coriolis force and therefore can be ignored.

6.4.2 The Geostrophic Balance

For $Ro_t << 1$, $Ro << 1$ and negligence of frictional effects, the Coriolis force and the horizontal pressure-gradient force are the only remaining large terms in the horizontal momentum equations to make up a force balance called the *geostrophic balance*.

6.4.3 Geostrophic Equations

The momentum equations for pure geostrophic flow are given by:

$$-f\,v_{\text{geo}} = -\frac{1}{\rho_o}\frac{\partial P}{\partial x} \tag{6.8}$$

$$+f\,u_{\text{geo}} = -\frac{1}{\rho_o}\frac{\partial P}{\partial y} \tag{6.9}$$

Accordingly, geostrophic flows run along lines of constant pressure, called *isobars*. With inclusion of the hydrostatic balance, which is valid for shallow-water processes, the latter equations can be formulated as:

$$\frac{\partial v_{\text{geo}}}{\partial z} = +\frac{g}{\rho f}\frac{\partial \rho}{\partial x} \tag{6.10}$$

$$\frac{\partial u_{\text{geo}}}{\partial z} = -\frac{g}{\rho f}\frac{\partial \rho}{\partial y} \tag{6.11}$$

These relations are known as the *thermal-wind equations*. According to these equations, the speed of geostrophic flow changes vertically in the presence of lateral density gradients. In oceanography, application of the thermal-wind equations is called the *geostrophic method*, being commonly used to derive the relative geostrophic flow field from measurements of density. See Pond and Pickard (1983) for a detailed description of the geostrophic method.

For an ocean uniform in density, the geostrophic balance reads:

$$-f v_{\text{geo}} = -g \frac{\partial \eta}{\partial x} \tag{6.12}$$

$$+f u_{\text{geo}} = -g \frac{\partial \eta}{\partial y} \tag{6.13}$$

The resultant geostrophic flow runs along lines of constant pressure, provided by sea-level elevations, and is independent of depth. Sea-level contours are therefore the *streamlines* of surface geostrophic flow. Such barotropic flow cannot produce much horizontal divergence and therefore tends to follow bathymetric contours (see Cushman-Roisin (1994)). Inspection of bathymetry maps provides first hints on the likely path of geostrophic currents! The geostrophic circulation around a low-pressure centre is referred to as *cyclonic*, whereas the circulation around a high-pressure centre is called *anticyclonic*.

Horizontal divergence of geostrophic flow is given by:

$$\frac{\partial u_{\text{geo}}}{\partial x} + \frac{\partial v_{\text{geo}}}{\partial y} = -\frac{\beta}{f} v_{\text{geo}} \tag{6.14}$$

where β is the meridional variation of the Coriolis parameter. This flow divergence/convergence occurs for equatorward or poleward flow and it can be typically ignored in regional studies on spatial scales <100 km. On larger scales, however, flow divergence associated with the beta effect is an important contributor to the steady-state wind-driven circulation in the ocean, being discussed in Sect. 6.9.

6.4.4 Vorticity

Vorticity is the ability of a flow to produce rotation. Imagine you throw a stick into the sea. If this imaginary stick starts to spin around, there must be some non-zero vorticity! A useful dynamical statement – *conservation of potential vorticity* – can be derived from consideration of the equations governing the dynamics of depth-independent, nonfrictional horizontal flows given by:

$$\frac{\partial u}{\partial t} + u \frac{\partial u}{\partial x} + v \frac{\partial u}{\partial y} - f v = -g \frac{\partial \eta}{\partial x} \tag{6.15}$$

$$\frac{\partial v}{\partial t} + u \frac{\partial v}{\partial x} + v \frac{\partial v}{\partial y} + f u = -g \frac{\partial \eta}{\partial y} \tag{6.16}$$

$$\frac{\partial \eta}{\partial t} + \frac{\partial (uh)}{\partial x} + \frac{\partial (vh)}{\partial y} = 0 \tag{6.17}$$

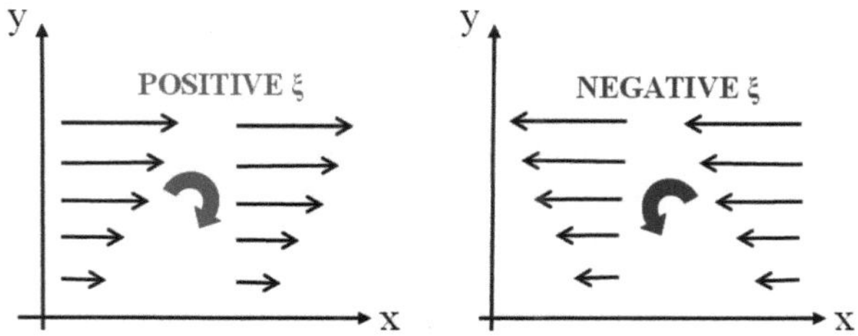

Fig. 6.2 Examples of horizontal flow fields exhibiting positive or negative relative vorticity

On the f-plane (f = constant), a combination of the momentum equations (6.15) and (6.16) yields:

$$\frac{\partial \xi}{\partial t} + u\frac{\partial \xi}{\partial x} + v\frac{\partial \xi}{\partial y} = -(f+\xi)\left(\frac{\partial u}{\partial x} + \frac{\partial v}{\partial y}\right) \tag{6.18}$$

where *relative vorticity* ξ (usually denoted by the Greek letter "xi") is defined by:

$$\xi = \frac{\partial v}{\partial x} - \frac{\partial u}{\partial y} \tag{6.19}$$

Figure 6.2 shows examples of horizontal flow fields exhibiting either positive or negative relative vorticity. Alternatively, Eq. (6.18) can be written in Lagrangian form as:

$$\frac{d\xi}{dt} = -(f+\xi)\left(\frac{\partial u}{\partial x} + \frac{\partial v}{\partial y}\right) \tag{6.20}$$

where the "d" symbol refers to a temporal change along the trajectory of the flow.

On the beta plane ($f = f_o + \beta y$), on the other hand, the equation for relative vorticity can be written as:

$$\frac{d(\xi + f)}{dt} = -(f+\xi)\left(\frac{\partial u}{\partial x} + \frac{\partial v}{\partial y}\right) \tag{6.21}$$

The only additional term appearing in this equation is $df/dt = \beta v$ associated with convergence/divergence inherent with meridional flow on the beta plane (as described by Eq. 6.14).

The Lagrangian version of the vertically integrated continuity equation (6.17) reads:

$$\frac{dh}{dt} = -h\left(\frac{\partial u}{\partial x} + \frac{\partial v}{\partial y}\right) \tag{6.22}$$

Accordingly, divergence/convergence of lateral flow experienced along the pathway will change the thickness of the water column and also produce relative vorticity. Equations (6.21) and (6.22) can be combined to yield:

$$\frac{d(PV)}{dt} = 0 \tag{6.23}$$

where the quantity PV, called *potential vorticity*, is defined by:

$$PV = \frac{f + \xi}{h} \tag{6.24}$$

In the context of vorticity, the Coriolis parameter (or inertial frequency) f is called *planetary vorticity*, and $f + \varsigma$ is called *absolute vorticity*.

6.4.5 Conservation of Potential Vorticity

The conservation principle of potential vorticity (6.23) in the ocean is akin to that of angular momentum for an isolated system. The best example is that of a ballerina spinning on her toes. With her arms stretched out, she spins slowly, but with her arms close to her body, she spins more rapidly. The important difference to the ballerina example is that a water column being initially at rest already exhibits potential vorticity owing to the rotation of Earth. Vertical squashing or stretching of this water column will produce relative vorticity and motion will appear (Fig. 6.3).

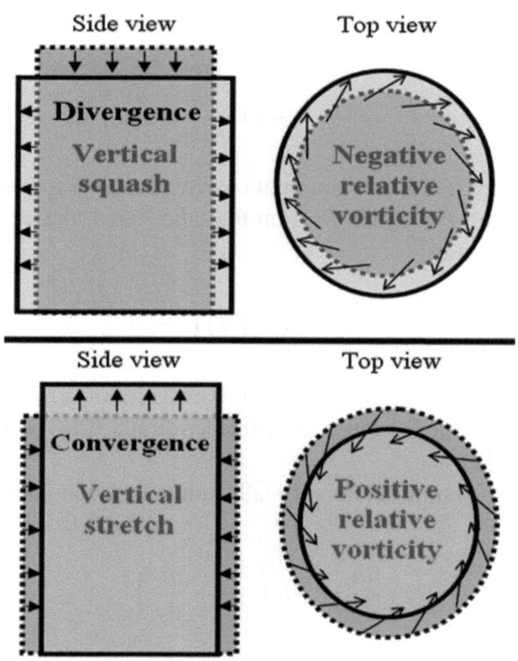

Fig. 6.3 Change of absolute vorticity associated with convergence or divergence of lateral flow (both for the northern hemisphere)

In a multi-layer non-frictional ocean, it can be shown (see Cushman-Roisin (1994)) that the conservation principle of potential vorticity applies to each layer separately; that is,

$$\frac{d(PV_i)}{dt} = 0$$

where i is the layer index, and:

$$PV_i = \frac{f + \xi_i}{h_i}$$

is the potential vorticity of a layer.

6.4.6 Topographic Steering

The ratio between relative vorticity and planetary vorticity scales as the Rossby number; that is;

$$\frac{\xi}{f} \approx \frac{U/L}{f} = \frac{U}{fL} = Ro \tag{6.25}$$

Quasi-geostrophic flows are flows characterised by a small Rossby number $Ro << 1$. For such flows, the conservation statement for potential vorticity turns into:

$$\frac{d(PV)}{dt} \approx \frac{d(f/h)}{dt} = 0 \Rightarrow \frac{f}{h} = \text{constant} \tag{6.26}$$

On the f-plane, this relation suggests that steady-state flows tend to follow bathymetric contours, a feature being referred to as *topographic steering*.

6.4.7 Rossby Waves

Relative vorticity is created by moving the water column to a different geographical latitude or by stretching or shrinking the water column through divergence/convergence of lateral flow. Waves created by disturbances of f are called *planetary Rossby waves*. Waves associated with disturbances of the thickness of the water column are referred to as *topographic Rossby waves*.

It can be shown that the dispersion relation of topographic Rossby waves in a fluid of uniform density is given by (Cushman-Roisin, 1994):

$$T = \frac{f\lambda_x}{\alpha g}\left[1 + (2\pi)^2 (R/\lambda)^2\right] \tag{6.27}$$

where T is wave period, λ is the true wavelength, λ_x is the apparent wavelength measured along bathymetric contours, α is the bottom slope, and R is the Rossby radius of deformation, given by (6.6). Consequently, the phase speed of wave propagation along topographic contours is given by:

$$c_x = \frac{\lambda_x}{T} = \frac{\alpha g}{f\left[1 + (2\pi)^2 (R/\lambda)^2\right]} \tag{6.28}$$

which implies that topographic Rossby waves propagate with shallower water on their right (left) in the northern (southern) hemisphere.

The following example gives an estimate of the phase speed of these waves. The deformation radius is $R = 100\,\text{km}$ for a depth of 100 m at mid-latitudes ($f = 10^{-4}\,\text{s}^{-1}$). Given a wavelength of $\lambda = 10\,\text{km}$ and a bottom slope of $\alpha = 0.01$ (corresponding to bathymetric variation of 10 m over 1 km), the phase speed of topographic Rossby waves is about 0.25 m/s or 22 km per day.

Planetary Rossby waves in a fluid of uniform density follow the dispersion relation (Cushman-Roisin, 1994):

$$T = \frac{\lambda_x}{\beta R^2}\left[1 + (2\pi)^2 (R/\lambda)^2\right] \tag{6.29}$$

where λ_x is the apparent wavelength measured in the zonal direction. Here, the zonal phase speed of wave propagation is given by:

$$c_x = -\frac{\beta R^2}{\left[1 + (2\pi)^2 (R/\lambda)^2\right]} \tag{6.30}$$

This zonal phase speed is always negative, implying a phase propagation with a westward component. The deformation radius is $R = 2200\,\text{km}$ for a deep-ocean depth of 5000 m at mid latitudes ($f = 10^{-4}\,\text{s}^{-1}$). With a wavelength of $\lambda = 100\,\text{km}$ and $\beta = 2.2 \times 10^{-11}\,\text{m}^{-1}\,\text{s}^{-1}$, we yield a phase speed of 5.5 mm/s corresponding to a distance of 175 km per year. Hence, planetary Rossby waves usually propagate at a much slower speed compared with topographic Rossby waves found predominantly at continental margins.

For relatively short waves, $\lambda << R$, the latter equation reduces to:

$$c_x = -\frac{\beta \lambda^2}{(2\pi)^2} \tag{6.31}$$

which implies that the zonal phase speed increases for larger wavelengths.

6.5 Exercise 16: Topographic Steering

6.5.1 Aim

The aim of this exercise is to explore the dynamics of barotropic quasi-geostrophic flow encountering variable bottom topography.

6.5.2 Model Equations

Consider an initially uniform zonal geostrophic flow U_{geo} that encounters a variable bottom topography. Since this background flow is uniform, we can predict the dynamics relative to this ambient flow from the horizontal momentum equations:

$$\frac{\partial u}{\partial t} + (u + U_{geo})\frac{\partial u}{\partial x} + v\frac{\partial u}{\partial y} - f v = -g\frac{\partial \eta}{\partial x} \tag{6.32}$$

$$\frac{\partial v}{\partial t} + (u + U_{geo})\frac{\partial v}{\partial x} + v\frac{\partial v}{\partial y} + f u = -g\frac{\partial \eta}{\partial y} \tag{6.33}$$

where η is a sea-level anomaly with reference to that driving the ambient geostrophic flow. The true flow has a velocity of $(U_{geo} + u, v)$. The vertically integrated continuity equation turns into:

$$\frac{\partial \eta}{\partial t} + \frac{\partial (uh)}{\partial x} + U_{geo}\frac{\partial h}{\partial x} + \frac{\partial (vh)}{\partial y} = 0 \tag{6.34}$$

Forcing appears in the continuity equation and is provided by interaction of the ambient geostrophic flow with variable bottom topography.

6.5.3 Task Description

Figure 6.4 shows the bathymetry used in this exercise. The model domain has a length of 150 km and a width of 50 km, resolved by lateral grid spacings of $\Delta x = \Delta y = 1$ km. The time step is set to $\Delta t = 20$ s. The ambient seafloor slopes downward in the y direction at a rate of 1 m per 1 km. The deepest part of the model domain is 100 m. The incident geostrophic flow of speed has to negotiate a bottom escarpment of 10 m in height variation over a distance of $W = 10$ km. The speed of the ambient geostrophic flow is set to $U_{geo} = +0.1$ m/s. All lateral boundaries are open.

Two scenarios are considered. The first scenario uses a Coriolis parameter of $f = -1 \times 10^{-4}\,\text{s}^{-1}$ (southern hemisphere), whereas the second scenario has $f = 1 \times 10^{-4}\,\text{s}^{-1}$ (northern hemisphere). A pseudo Rossby number can be constructed on the basis of ambient parameters yielding $Ro = U_{geo}/(W\,|f|) = 0.1$ for both scenarios. This number, however, is not a true Rossby number, since it is not based

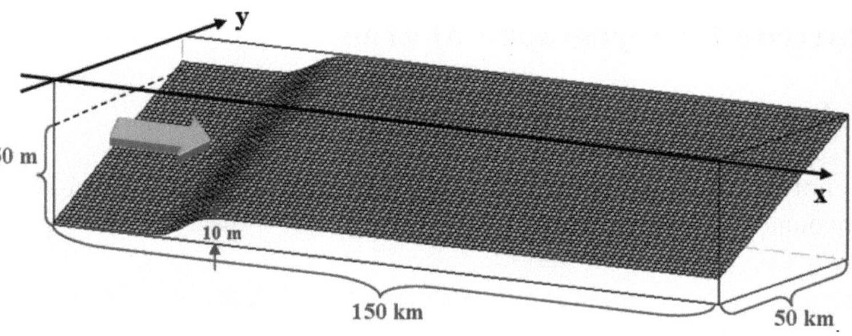

Fig. 6.4 Bathymetry for Exercise 16

on the velocity scale and lengthscale of dynamical perturbations that develop in interaction with variable bathymetry.

The total simulation time of experiments is 20 days with data outputs at every 6 h. A narrow source of Eulerian tracer concentration of unity is prescribed at the western boundary to visualise the structure of the flow. Wind-stress forcing and lateral momentum diffusion are ignored. Zero-gradient conditions are employed for all variables at open boundaries. Additional smoothing algorithms are implemented near the western and eastern open boundaries to avoid reflection of topographic Rossby waves.

6.5.4 Caution

The first-order Shapiro filter does not work well for processes dominated by the geostrophic balance. This filter operates to gradually decrease sea-level gradients, hence diminishing the barotropic horizontal pressure-gradient force that is the principal driver of geostrophic flows in the ocean. For this reason, the Shapiro filter is disabled in this and most of the subsequent model applications, if not stated otherwise.

6.5.5 Sample Code

The folder "Exercise 16" of the CD-ROM contains the computer codes for this exercise. The file "info.txt" gives additional information.

6.5.6 Results

In Scenario 1, the flow largely follows bathymetric contours and the topographic steering mechanism appears to work (Fig. 6.5). Given that the flow enters the model domain through the upstream boundary with zero relative vorticity, the conservation principle of potential vorticity (6.23) has the solution:

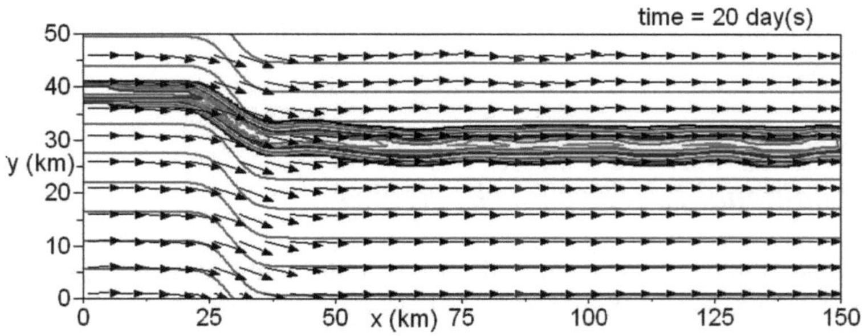

Fig. 6.5 Exercise 16. Scenario 1. Snapshot of flow field (*arrows*, averaged over 5×5 grid cells) and Eulerian tracer concentration (*crowded lines*) after 20 days of simulation. Bathymetric contours are overlaid

$$\xi = f\left(\frac{h_o + \Delta h}{h_o} - 1\right) \tag{6.35}$$

where h_o is the initial thickness of the water column, and Δh is the change in thickness of the water column along the flow trajectory.

In Scenario 1, water-column squeezing over the bottom escarpment leads to a flow whose relative vorticity matches the curvature of bathymetric contours. The propagation direction of topographic Rossby waves is the same as that of the ambient flow, so that these waves propagate rapidly away from their generation zone.

Surprisingly, something different happens in Scenario 2 (Fig. 6.6). Here, water-column squeezing over the bottom escarpment creates relative vorticity of opposite sign to that of Scenario 1. In response to this, the flow crosses bathymetric contours into deeper water. This initiates a standing topographic Rossby wave of a wavelength such that its phase speed (given by Eq. 6.28) is compensated by the speed of the ambient flow. For the configuration of this exercise, the resultant wave pattern attains a horizontal amplitude of 20 km and a wavelength of 50 km. Obviously, situations in which the ambient flow runs opposite to the propagation direction of topographic Rossby waves support the creation of such standing waves. Despite the

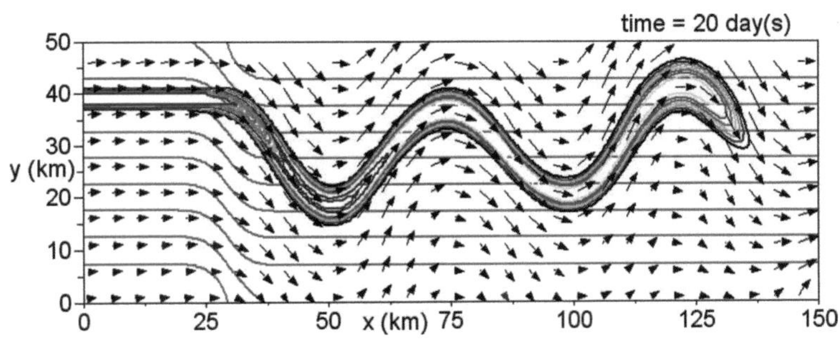

Fig. 6.6 Exercise 16. Same as Fig. 6.5, but for Scenario 2

appearance of these waves, the flow still tends to follow bathymetric contours, at least, on average.

6.5.7 Additional Exercise for the Reader

Repeat this exercise with a reduced speed of the ambient geostrophic flow of $U_{geo} = 0.05$ m/s. This setting corresponds to a pseudo Rossby number, based on the width of the bottom-escarpment zone, of $Ro = 0.05$, which is half that used before. Explore the topographic steering mechanism for this modified situation and verify whether the wavelength of standing topographic Rossby waves increases or decreases. Because of weaker ambient flow, the reader should double the total simulation time.

6.6 Instability of Lateral Shear Flows

6.6.1 Theory

Under certain conditions, lateral shear flows in a homogeneous fluid become dynamically unstable, start to meander and eventually break up into turbulent eddies. This process is called *barotropic instability*. Consider a purely zonal frictionless flow $\overline{u}(y)$ on the beta plane of a speed that varies in the transverse direction, such as the flow shown in Fig. 6.7. In the absence of disturbances and in a steady state, this flow is perfectly in geostrophic balance; that is,

$$f\overline{u} = -g\frac{d\overline{\eta}}{dy}$$

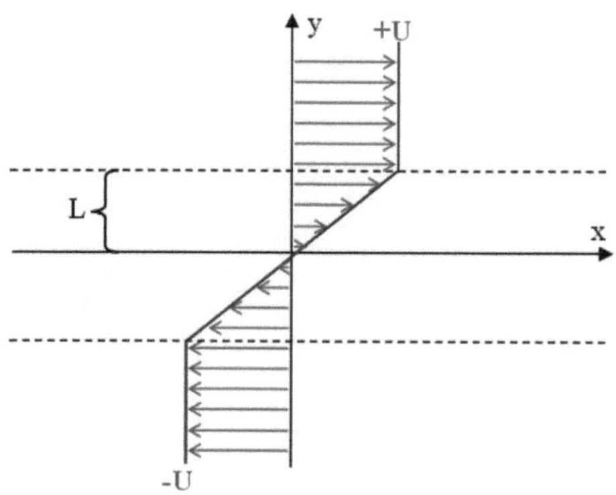

Fig. 6.7 A shear-flow profile on the f plane used in Exercise 17 that is unstable to long waves

With addition of small perturbations; that is $u = \overline{u} + u'$, $v = v'$, and $\eta = \overline{\eta} + \eta'$, the momentum equations can be written as (see Cushman-Roisin, 1994):

$$\frac{\partial u'}{\partial t} + \overline{u}\frac{\partial u'}{\partial x} + v'\frac{\partial \overline{u}}{\partial y} - fv' = -g\frac{\partial \eta'}{\partial x}$$

$$\frac{\partial v'}{\partial t} + \overline{u}\frac{\partial v'}{\partial x} + fu' = -g\frac{\partial \eta'}{\partial y}$$

Based on scaling arguments and with a focus on the initial appearance of disturbances, the smallest terms such as $u'\partial u'/\partial x$ have been dropped in the above equations. In addition to this, vertical velocity is assumed negligibly small, so that the continuity equation can be expressed as:

$$\frac{\partial u'}{\partial x} + \frac{\partial v'}{\partial y} = 0$$

Accordingly, sea surface pressure is not directly calculated from these equations, but rather appears implicitly as a requirement to produce a horizontal flow void of lateral divergence/convergence. The flow follows the contours of a certain streamfunction, called streamlines. Here, a streamfunction can be constructed from:

$$u' = -\frac{\partial \psi}{\partial y} \quad \text{and} \quad v' = +\frac{\partial \psi}{\partial x}$$

which satisfies the previous continuity equation. By use of this streamfunction, the linearised equations can be combined to yield a single equation for the streamfunction:

$$\left(\frac{\partial}{\partial t} + \overline{u}\frac{\partial}{\partial x}\right)\nabla^2\psi + \left(\beta - \frac{d^2\overline{u}}{\partial y^2}\right)\frac{\partial \psi}{\partial x} = 0$$

where β is the meridional variation of the Coriolis parameter. The solution of this equation depends on the specific form of $\overline{u}$. It can be shown that a necessary condition for barotropic instability to occur is that the function:

$$\beta - \frac{d^2\overline{u}}{dy^2} \tag{6.36}$$

vanishes at least once within the model domain. This result was first derived by Kuo (1949).

6.6.2 Instability to Long Waves

On the f-plane ($\beta = 0$), the shear-flow profile shown in Fig. 6.7 satisfies the conditions necessary for instability to develop. It can be shown that initial disturbances of a wavelength greater than 9.8 L, where L is the half-width of the shear zone, are subtle to instability and grow rapidly. This wave does not travel but amplifies with time. Disturbances of a shorter wavelength travel with the flow without growth (Cushman-Roisin, 1994). Hence, the barotropic instability process discriminates disturbances according to their wavelength.

6.7 Exercise 17: Barotropic Instability

6.7.1 Aim

The aim of this exercise is to simulate dynamic instabilities produced by horizontal shear flows.

6.7.2 Model Equations

Under the assumption of a steady zonal geostrophic background flow, U_{geo}, the equations governing the problem can be written as:

$$\frac{\partial u}{\partial t} + (u + U_{geo})\frac{\partial u}{\partial x} + v\frac{\partial (u + U_{geo})}{\partial y} - fv = -g\frac{\partial \eta}{\partial x} \tag{6.37}$$

$$\frac{\partial v}{\partial t} + (u + U_{geo})\frac{\partial v}{\partial x} + v\frac{\partial v}{\partial y} + fu = -g\frac{\partial \eta}{\partial y} \tag{6.38}$$

$$\frac{\partial \eta}{\partial t} + \frac{\partial (uh)}{\partial x} + U_{geo}\frac{\partial h}{\partial x} + \frac{\partial (vh)}{\partial y} = 0 \tag{6.39}$$

where u, v and η are flow and sea level disturbances with respect to the ambient flow. These equations are identical to those in Exercise 16, with the addition that the geostrophic background flow is allowed to vary in the y-direction. The reason why the full equations rather than simplified equations of the previous section are used here is that we want to be able to simulate the entire instability process and not only its initial phase.

6.7.3 Task Description

The model domain of this exercise is an open channel of 10 km in length and 5 km in width, bounded by coasts along the northern and southern boundaries. Lateral grid spacings are set to $\Delta x = \Delta y = 100$ m. The time step is $\Delta t = 3$ s. Open boundaries are treated as cyclic boundaries. Total water depth is set to a uniform value of 10 m. Random disturbances of 0.1 m in amplitude are added to support the onset of dynamical instabilities.

Wind-stress forcing, lateral momentum diffusion and bottom friction are disabled. The Coriolis parameter is set to a mid-latitude value of $f = 1 \times 10^{-4}\ \mathrm{s}^{-1}$ (northern hemisphere). The model is forced by prescription of an ambient geostrophic lateral shear flow of a speed of $U = \pm 0.2$ m/s with a shear zone of 800 m in width (see Fig. 6.7).

The barotropic instability mechanism is visualised by addition of Eulerian tracer concentration to the model domain with values of unity in the northern half of the channel and zero values in the southern half. The total simulation time is 36 h with outputs of variables at every 30 min.

6.7.4 Results

The shallow-water model is able to simulate the barotropic instability process. Instabilities appear after 18 h of simulation on wavelengths of 5 km (Fig. 6.8). This exceeds fivefold the width of the shear zone and therefore agrees with theory. The limited size of the model domain and the use of cyclic boundary conditions, however, modify the wavelength such that the wave pattern fits into the model domain. Disturbances grow rapidly with time. After 1 day of simulation, wave disturbances appear to break and form clockwise vortices of about 3 km in diameter (Fig. 6.9). These eddies induce vigorous horizontal stirring.

The Earth's rotation does not play a role in the perturbations simulated here that are characterised by values of both the temporal Rossby number and the Rossby

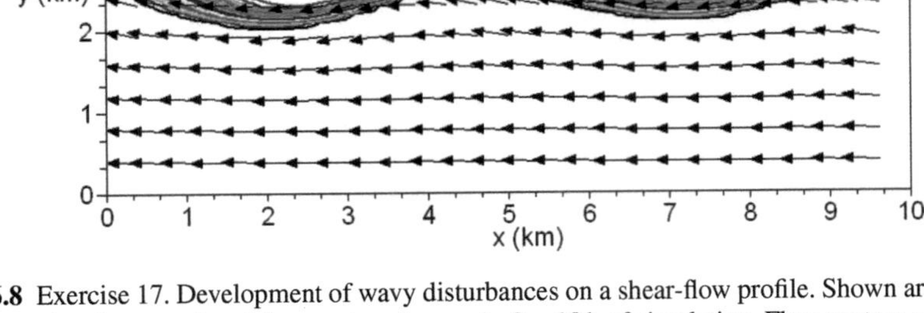

Fig. 6.8 Exercise 17. Development of wavy disturbances on a shear-flow profile. Shown are tracer concentration (*contours*) and flow vectors (*arrows*) after 18 h of simulation. Flow vectors are averaged over 3×3 grid cells

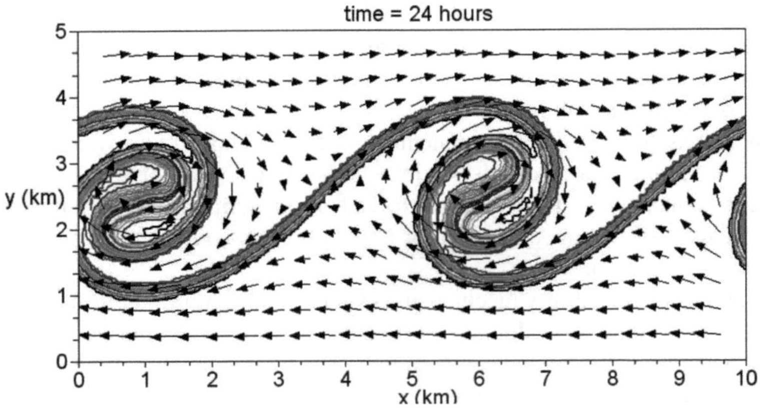

Fig. 6.9 Exercise 17. Same as Fig. 6.8, but after 24 h of simulation

number exceed unity in this application, so that the Earth's rotation does not influence the perturbations simulated here. It is worth mentioning that similar dynamics will result with use of the shear flow as initial condition rather than background flow. The reader is encouraged to verify the latter statement. A much larger model domain would be required to avoid influences of the open and closed boundaries.

6.7.5 Sample Code and Animation Script

The folder "Exercise 17" on the CD-ROM contains the computer codes. The file "info.txt" contains additional information.

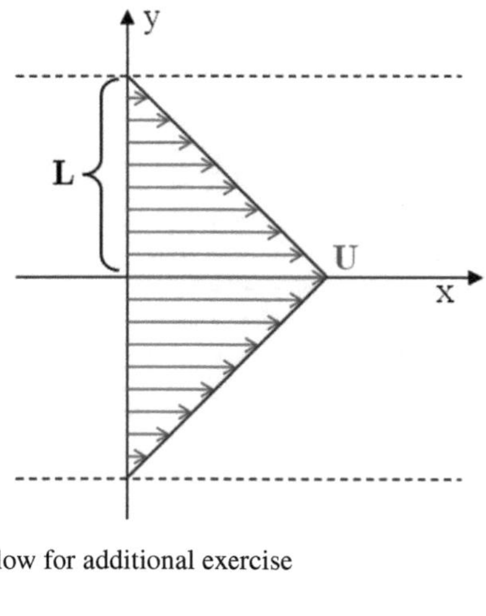

Fig. 6.10 Ambient shear flow for additional exercise

6.7.6 Additional Exercise for the Reader

Explore whether the ambient flow field shown in Fig. 6.10 is subtle to the barotropic instability process. Use $L = 500$ m and $U = 0.2$ m/s together with a constant Coriolis parameter.

6.8 The Wind-Driven Circulation of the Ocean

6.8.1 The Dynamical Structure of the Ocean

Frictional effects are only significant in thin frictional boundary layers, called *Ekman layers*, establishing near the sea surface and the seafloor of the ocean and along coastlines. Larger-scale flow in the ocean interior is essentially void if friction and is therefore governed by a balance between the horizontal pressure-gradient force and the Coriolis force – the geostrophic balance. The ocean's dynamics can therefore be classified by two distinct dynamic regimes: frictional boundary layers and the geostrophic interior (Fig. 6.11).

6.8.2 Steady-State Dynamics and Volume Transport

Steady-state dynamics reflect a state in which the surface pressure field that drives geostrophic flow has reached an equilibrium. This equilibrium implies that this flow, when averaged over the entire water column, is void of lateral divergence/convergence. This can be expressed by the balance:

$$\frac{\partial Q_x}{\partial x} + \frac{\partial Q_y}{\partial y} = 0 \tag{6.40}$$

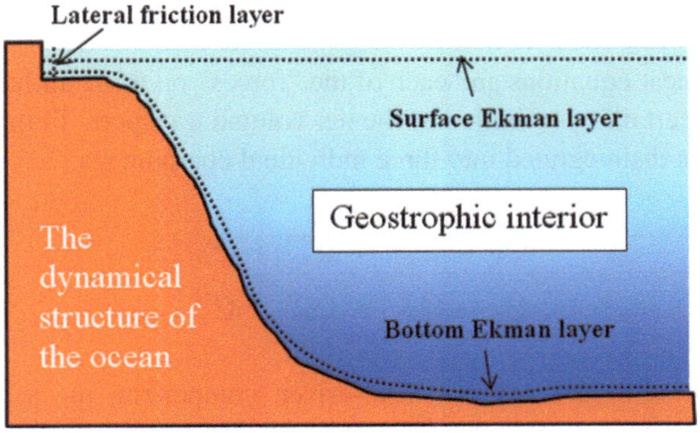

Fig. 6.11 The dynamical structure of the ocean

where Q_x and Q_y are components of a vector called *volume transport* that is yielded from vertical integration of the lateral flow over the entire water column. Volume transport carries units of m^3/s per unit width of the flow. In recognition of Harald Sverdrup's work (e.g., Sverdrup, 1947), oceanographers often express volume transports in units of Sverdrups (Sv) with 1 Sv being equivalent to 1×10^6 m^3/s. The net volume transport is composed of several contributions that compensate each other in order to achieve a steady-state sea level. The details of this compensation is discussed in the following.

6.8.3 A Simplified Model of the Wind-driven Circulation

A simplified model of the wind-driven circulation can be constructed when considering an ocean of uniform water depth h_o being void of density stratification. Large-scale oceanic flows are associated with very small Rossby numbers. The nonlinear terms can therefore be ignored. For simplicity, we also neglect horizontal momentum diffusion. Accordingly, the vertically integrated momentum equations (6.1) and (6.2) can be written as:

$$\frac{\partial Q_x}{\partial t} - f Q_y = -g h_o \frac{\partial \eta}{\partial x} + \frac{\tau_x^{wind} - \tau_x^{bot}}{\rho_o}$$

$$\frac{\partial Q_y}{\partial t} + f Q_x = -g h_o \frac{\partial \eta}{\partial y} + \frac{\tau_y^{wind} - \tau_y^{bot}}{\rho_o}$$

In a steady state, the latter equations turn into:

$$-f Q_y = -g h_o \frac{\partial \eta}{\partial x} + \frac{\tau_x^{wind} - \tau_x^{bot}}{\rho_o} \tag{6.41}$$

$$+f Q_x = -g h_o \frac{\partial \eta}{\partial y} + \frac{\tau_y^{wind} - \tau_y^{bot}}{\rho_o} \tag{6.42}$$

These are linear equations and each of the "forces" on the right-hand side can be attributed to a certain contribution to the net volume transport. To this end, volume transport can be disintegrated into three individual components:

$$Q_x = Q_x^{ek,s} + Q_x^{geo} + Q_x^{ek,b}$$

$$Q_y = Q_y^{ek,s} + Q_y^{geo} + Q_y^{ek,b}$$

where the first term denotes the wind-driven component, the second term the geostrophic component, and the last term the component attributed to bottom friction.

6.8.4 The Surface Ekman Layer

Winds impose a tangential frictional stress to the sea surface that transfers momentum into the ocean by means of vertical diffusion of momentum. This friction only plays a role in a surface layer of finite depth, the so-called *surface Ekman layer*. It can be shown (see Cushman-Roisin, 1994) that the thickness of this layer is given by:

$$\delta_{ek} = \sqrt{\frac{2A_z}{f}} \tag{6.43}$$

where A_z is vertical eddy viscosity and f is the Coriolis parameter. Typical thicknesses are 50–150 m with increasing values towards the equator. The vertical lengthscale D associated with the Ekman layer can also be derived from scaling considerations. The ratio between the friction force and the Coriolis force can be expressed by means of the so-called *Ekman number* given by:

$$Ek = \frac{A_z}{D^2 f}$$

The surface Ekman layer establishes on a lengthscale corresponding to $Ek \approx 1$, or:

$$D = \sqrt{\frac{A_z}{f}}$$

which is of the order of the right-hand-side of Eq. (6.43).

6.8.5 Ekman-layer Transport

According to Eqs. (6.41) and (6.42), the net volume transport in the surface Ekman layer (also called *Ekman drift*) is given by:

$$Q_x^{ek,s} = +\frac{\tau_y^{\text{wind}}}{f\,\rho_o} \quad \text{and} \quad Q_y^{ek,s} = -\frac{\tau_x^{\text{wind}}}{f\,\rho_o} \tag{6.44}$$

This Ekman drift, corresponding to the vertically-averaged flow in this layer, is directed at right angle with respect to the wind direction, 90° to the right in the northern hemisphere and 90° to the left in the southern hemisphere. Figure 6.12 shows an example of wind-induced Ekman transports at mid-latitudes in the northern hemisphere.

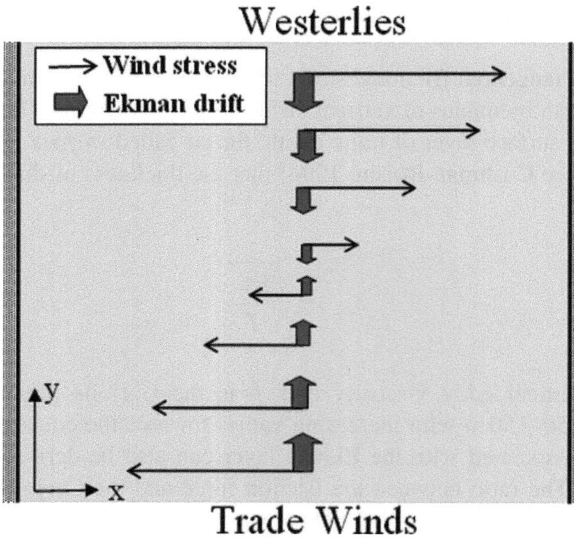

Fig. 6.12 Sketch of mean wind pattern and associated Ekman-layer transports at mid-latitudes in the northern hemisphere

6.8.6 Ekman Pumping

Wind stresses are the principle driver of the general geostrophic circulation in the ocean. The way by which wind stresses drive geostrophic flow is indirect via divergence or convergence of lateral flow in the surface Ekman layer. From (6.44), the latter can be specified as:

$$\frac{\partial Q_x^{ek,s}}{\partial x}+\frac{\partial Q_y^{ek,s}}{\partial y}=\frac{1}{\rho_o f}\left\{\frac{\partial \tau_y^{\text{wind}}}{\partial x}-\frac{\partial \tau_x^{\text{wind}}}{\partial y}\right\}+\frac{\beta}{f^2}\frac{\tau_x^{\text{wind}}}{\rho_o} \tag{6.45}$$

where the last term is small compared with the others and can be ignored.

The resultant stretching or squeezing of the surface Ekman layer is called *Ekman pumping*. In the absence of other processes, this Ekman pumping would initiate a rise or drop of the sea level. Lateral pressure gradients associated with resultant sea-level anomalies are the principle driver of deep-reaching geostrophic flow in the ocean. Ekman pumping is interpreted by some authors as a vertical velocity at the base of the surface mixed layer that leads to deformation of isopycnals in the ocean interior. This interpretation is misleading and incorrect given that flow divergence in a layer of incompressible fluid can only lift the surface level of this layer, but never its bottom level.

6.8.7 The Sverdrup Balance

On the β plane, the divergence of lateral volume transport inherent with geostrophic flow is given by:

$$\frac{\partial Q_x^{geo}}{\partial x} + \frac{\partial Q_y^{geo}}{\partial y} = -\frac{\beta}{f} Q_y^{geo} \tag{6.46}$$

On the large scale, this divergence (or convergence) of the geostrophic flow balances the convergence (or divergence) of the drift in the surface Ekman layer. Effects associated with bottom friction are irrelevant here. Under the assumption of purely zonal wind ($\tau_y^{wind} = 0$), the steady-state balance leading to an equilibrium sea-level distribution is given by:

$$\beta Q_y^{geo} \approx -\frac{1}{\rho_o} \frac{\partial \tau_x^{wind}}{\partial y} \tag{6.47}$$

This balance is called the *Sverdrup relation* (Sverdrup, 1947) and allows for calculation of the meridional geostrophic volume transport (also called *Sverdrup transport*) from knowledge of the average zonal wind-stress distribution. The corresponding zonal volume transport can be estimated from:

$$\frac{\partial Q_x^{geo}}{\partial x} + \frac{\partial Q_y^{geo}}{\partial y} \approx 0 \tag{6.48}$$

where the β effect can be ignored since this equation is used for diagnostic purposes only. It is important to note that the Sverdrup balance can only establish with existence of a meridional boundary. The dynamics of unbounded flows, such as the *Antarctic Circumpolar Current*, is more complex.

6.8.8 Interpretation of the Sverdrup Relation

First and foremost, the Sverdrup relation implies that latitudes of vanishing wind-stress curl ($\partial \tau_x^{wind}/\partial y = 0$) coincide with regions of vanishing meridional geostrophic flow. Hence, these regions form natural boundaries that, for instance, separate subtropical from subpolar gyres in the ocean.

In the midlatitude ocean of the northern hemisphere, the main wind pattern consists of trades to the south and westerlies to the north. This wind pattern provides $\partial \tau_x^{wind}/\partial y > 0$ and produces a convergence of volume transport in the surface Ekman layer. Hence, equatorward Sverdrup transport is required to balance this convergence.

Since no geostrophic flow is possible across the natural boundaries marked by the maximum trade winds and the maximum westerlies, this equatorward flow must

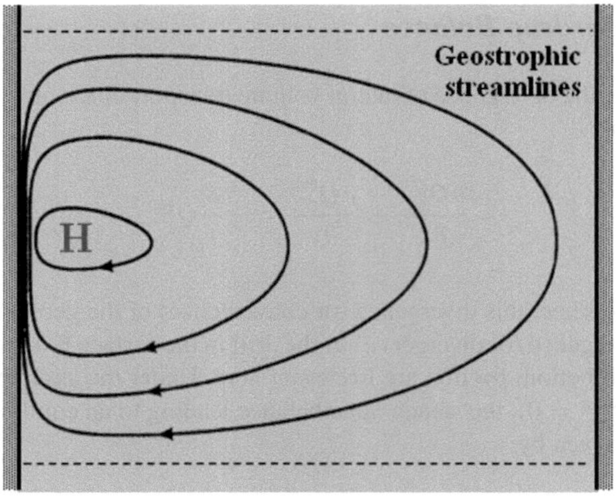

Fig. 6.13 Sketch of geostrophic circulation at mid-latitudes in the northern hemisphere. Sea-level contours are the streamlines of barotropic geostrophic flows

be compensated by a poleward return flow. This return flow occurs in a narrow zone along the western boundary in which the Sverdrup relation loses its validity. The wind-driven geostrophic circulation takes the form of an asymmetric gyre, with a slow equatorward flow occupying most of the domain and swift boundary-layer current on the western side returning water masses northward (Fig. 6.13).

6.8.9 The Bottom Ekman Layer

Ekman layers, 10–25 m in thickness, can establish in vicinity of the seafloor. For simplicity, we can approximate bottom friction by a linear bottom-drag law, given by:

$$\frac{\tau_x^{\mathrm{bot}}}{\rho_o} = r\frac{Q_x^{\mathrm{geo}}}{h_o} \quad \text{and} \quad \frac{\tau_y^{\mathrm{bot}}}{\rho_o} = r\frac{Q_y^{\mathrm{geo}}}{h_o} \tag{6.49}$$

where r is a friction parameter carrying units of m/s. According to (6.41) and (6.42), the resultant divergence of lateral volume transport in the bottom Ekman layer is given by:

$$\frac{\partial Q_x^{ek,b}}{\partial x} + \frac{\partial Q_y^{ek,b}}{\partial y} = \frac{r}{h_o\, f}\left(\frac{\partial Q_y^{\mathrm{geo}}}{\partial x} - \frac{\partial Q_x^{\mathrm{geo}}}{\partial y}\right) - \frac{\beta}{f}\frac{r}{h_o}Q_y^{\mathrm{geo}} \tag{6.50}$$

where the last term is negligibly small compared with the other terms. The latter equation implies that it is the relative vorticity of the geostrophic flow that produces a flow convergence/divergence in the bottom Ekman layer. For consistency with the analytical solution of the Ekman-layer equations (see Cushman-Roisin, 1994), the linear friction parameter has to be chosen according to:

$$r = 0.5\,\delta_{ek} f \tag{6.51}$$

where the Ekman-layer thickness is given by (6.43).

6.8.10 Western Boundary Currents

Western boundary currents are regions in which the Sverdrup relation is not valid and where frictional forces come into play. Western boundary currents are found at the western continental rise of all oceans. The mechanism that leads to these currents is called *westward intensification*, first described by Stommel (1948). The typical width of western boundary currents is 20–50 km and their speed can exceed 1 m/s. On these scales, the direct impact of wind-driven Ekman pumping can be ignored and the dynamical equations of our simplified model governing this regime are given by:

$$-\frac{\beta}{f} Q_y^{geo} = \frac{r}{h_o\, f}\left(\frac{\partial Q_y^{geo}}{\partial x} - \frac{\partial Q_x^{geo}}{\partial y}\right)$$

Since the velocity shear is much larger across the stream than along it; that is, $|\partial Q_y^{geo}/\partial x| >> |\partial Q_x^{geo}/\partial y|$, the latter equation can be approximated as:

$$\frac{\partial Q_y^{geo}}{\partial x} = -\alpha Q_y^{geo} \tag{6.52}$$

where $\alpha = \beta h_o/r$ or, with (6.51), $\alpha = 2\beta h_o/(f\delta_{ek})$. The solution of the latter equation is:

$$Q_y^{geo} = Q_o(y)\exp(-\alpha x) \tag{6.53}$$

where x is distance from the western coast, and $Q_o(y)$ is the maximal value of volume transport occuring at $x = 0$. This maximum can be derived from the conditions that the boundary solution has to match the Sverdrup solution outside the western boundary. Cushman-Roisin (1994) details the full mathematical procedure. Figure 6.14 shows the final structure of the meridional geostrophic flow component.

The width of the western boundary current can be estimated from the distance from the coast at which the volume transport according to (6.53) has decreased to fraction of $\exp(-\pi)$ (4.3%) of the coastal value. Using (6.53), this distance L is given by:

$$L = 0.5\pi \frac{f\delta_{ek}}{\beta h_o}$$

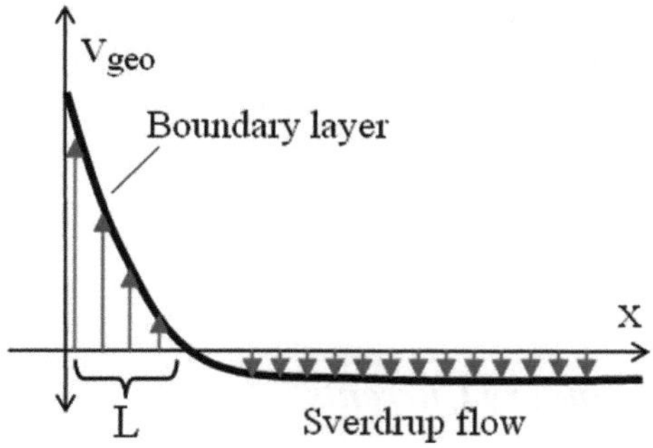

Fig. 6.14 Sketch of the structure of the meridional geostrophic flow component v_{geo}

Using typical values ($f = 1\times10^{-4}\text{s}^{-1}$, $\delta_{ek} = 10\,\text{m}$, $\beta = 2\times10^{-11}\text{m}^{-1}\,\text{s}^{-1}$, and $h_o = 4000\,\text{m}$), we yield $L \approx 20\,\text{km}$.

6.8.11 The Role of Lateral Momentum Diffusion

Bottom friction is irrelevant for the Sverdup regime that occupies most of the domain. Therefore, the Sverdup relation is also valied in a stratified ocean in which the flow does not extend to the seafloor. In this case, the above geostrophic volume transports represent the vertical integral of the baroclinic geostrophic flow and the vertical structure of this flow is irrelevant for the resultant steady-state surface pressure field.

In our simplified model, the western boundary current arises exclusively due to bottom friction and the model fails in the absence of near-bottom flows. The use of lateral momentum diffusion instead of bottom friction in the governing equations overcomes this problem and enables the analytical description of western boundary currents detached from the seafloor. The equations governing this problem are more complex and therefore not included in this book. The interested reader is referred to the work by Munk (1950).

6.9 Exercise 18: The Wind-Driven Circulation

6.9.1 Aim

The aim of this exercise is to reveal the wind-driven circulation of the ocean at midlatitudes consisting of subtropical gyres and western boundary currents.

6.9.2 Task Description

In this exercise, we apply the shallow-water equations to study the wind-driven circulation in a closed rectangular ocean basin of 1000 km in length and 500 km

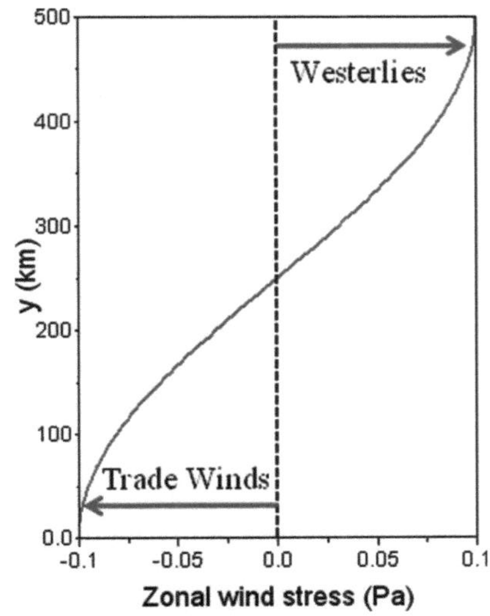

Fig. 6.15 Wind-stress forcing for Exercise 18

in width, resolved by a grid spacing of $\Delta x = \Delta y = 10\,\text{km}$. The depth of this basin is set to 1000 m and the time step is set to $\Delta t = 20\,\text{s}$. The ocean is assumed to be uniform in density.

The circulation is driven by a simplified zonal wind-stress forcing (Fig. 6.15) mimicking the general atmospheric wind pattern at mid-latitudes of the northern hemisphere. Indeed, the dimensions used are different from the real situation and serve for demonstration purposes only. The wind-stress field is slowly introduced over an adjustment period of 50 days to avoid appearance of initial disturbances. The total simulation time is 100 days with data outputs at every 2.5 days.

Three different scenarios are considered. In the first scenario, the Coriolis parameter is set to a constant value of $f = 1\times10^{-4}\ \text{s}^{-1}$. Lateral diffusion and the nonlinear terms are disabled. In this case, the wind-stress forcing produces a continuous Ekman pumping that can only be compensated by frictional effects in the entire model domain. To achieve reasonable current speeds, the bottom-friction parameter has to be set to an unrealistically high value of $r = 0.1$ m/s corresponding to an enormous bottom Ekman layer that, according to (6.51), extends the entire water column.

In the second scenario, the Coriolis parameter is assumed to vary with the meridional distance y according to the beta-plane approximation; that is, $f = f_o + \beta y$, where $f = 1\times10^{-4}\ \text{s}^{-1}$ at the southern boundary, y is distance to the north, and β is chosen at $= 4\times10^{-11}\ \text{m}^{-1}\ \text{s}^{-1}$. Note that β is twice the real value. A smaller value of $r = 0.01$ m/s is chosen, implying a bottom Ekman layer of 200 m in thickness, which overestimates the real situation by one order of magnitude. Both lateral momentum diffusion and the nonlinear terms are disabled.

The third scenario includes a variable Coriolis parameter, the nonlinear terms, lateral momentum diffusion with no-slip lateral boundary conditions (see Sect. 5.10) and uniform lateral eddy viscosity with a value of $A_h = 500\,\mathrm{m^2/s}$. Bottom friction is disabled.

6.9.3 Results

For a flat Earth (Scenario 1), the wind-stress forcing imposed creates a symmetrical clockwise oceanic gyre with elevated sea level in its high-pressure centre (Fig. 6.16). Recall that sea-level contours are the streamlines of surface geostrophic flow and that the spacing between adjacent contours is a measure of the speed of this flow. The apparent slight asymmetry of streamlines is caused by the sea-level effects in the divergence terms of the vertically integrated continuity (6.17).

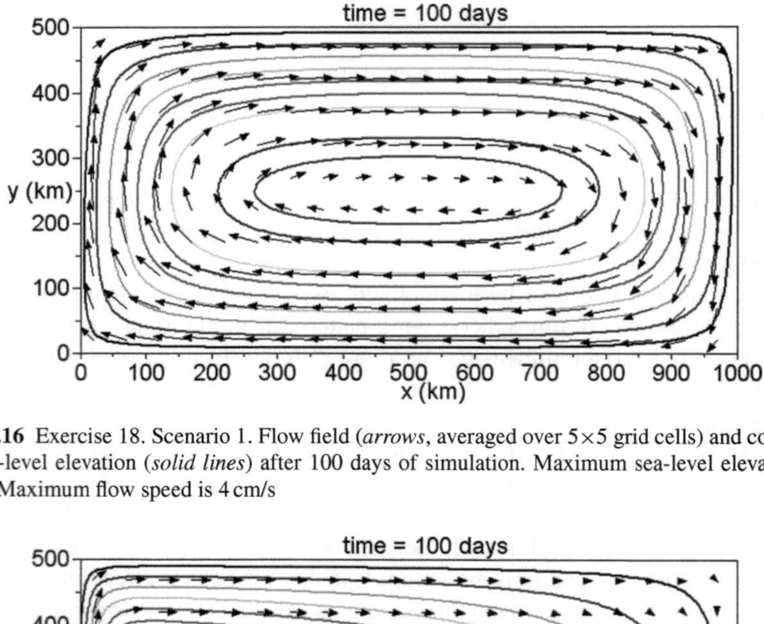

Fig. 6.16 Exercise 18. Scenario 1. Flow field (*arrows*, averaged over 5×5 grid cells) and contours of sea-level elevation (*solid lines*) after 100 days of simulation. Maximum sea-level elevation is 3 cm. Maximum flow speed is 4 cm/s

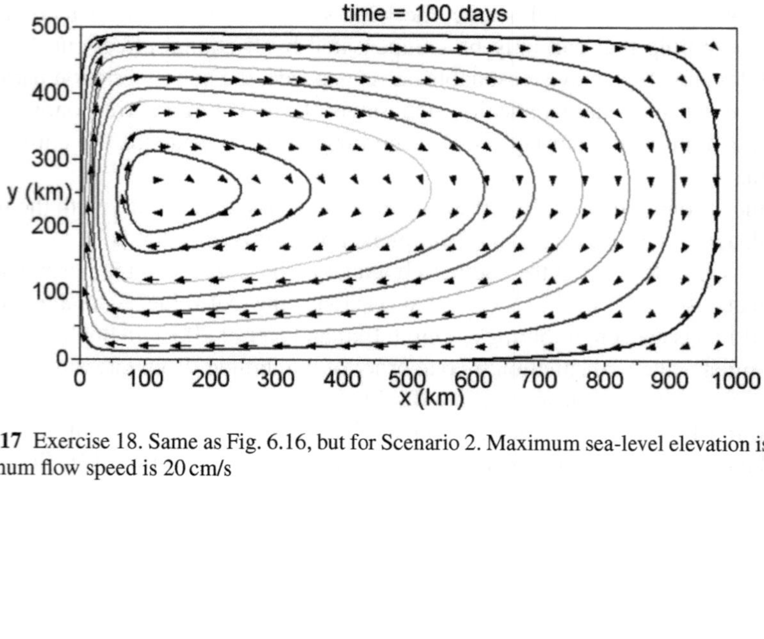

Fig. 6.17 Exercise 18. Same as Fig. 6.16, but for Scenario 2. Maximum sea-level elevation is 7 cm. Maximum flow speed is 20 cm/s

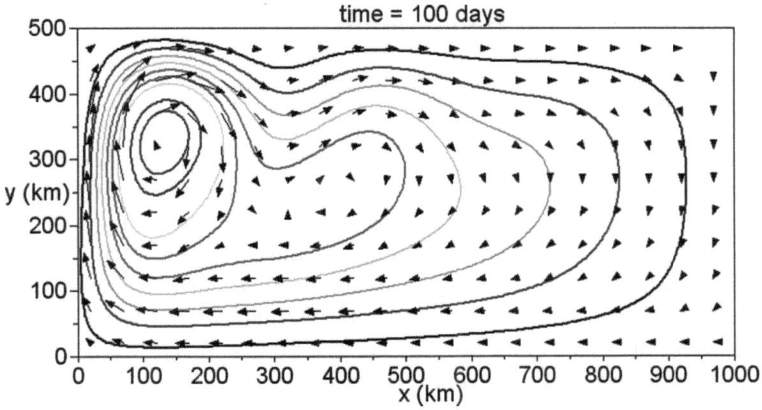

Fig. 6.18 Exercise 18. Same as Fig. 6.16, but for Scenario 3. The maximum sea-level elevation is 14 cm. The maximum flow speed is 20 cm/s

On a spherical Earth approximated by the β-plane approximation (Scenario 2), a swift western boundary current establishes (Fig. 6.17). Compared with Scenario 1, the high-pressure centre is moved closer to the western boundary. Hence, existence of western boundary currents is the definite scientific proof that the Earth is not flat.

With the inclusion of lateral momentum diffusion and coastal friction (instead of bottom friction) together with the nonlinear terms (Scenario 3), an equatorward countercurrent establishes next to the western boundary current (Fig. 6.18), first mathematically described by Munk (1950).

In contrast to our closed-basin case, real western boundary currents reach farther poleward due to inertia and can therefore trigger substantial poleward heat transports influencing climate in adjacent countries. Owing to this heat transport, northern Europe is on average 9° Celsius warmer than elsewhere for the same geographical latitude.

6.9.4 Sample Code and Animation Script

The folder "Exercise 18" on the CD-ROM contains the computer codes for this exercise. The file "info.txt" gives additional information.

6.9.5 Additional Exercises for the Reader

Using the bathymetry creator of previous exercises, include a mid-ocean ridge to the bathymetry and explore resultant changes in the dynamical response of the ocean. Explore changes in the circulation for different values of lateral eddy viscosity.

6.10 Exercise 19: Baroclinic Compensation

6.10.1 Background

Adjustment toward a steady state is in our previous model of the wind-driven mid-latitude circulation only possible if flow convergence in the upper ocean is compensated by flow divergence in deeper layers of the ocean. Whereas the convergence of Ekman drift leads to establishment of a centre of elevated sea level, it is obvious the flow divergence in the ocean interior leads to downward displacements of density interfaces. Hence, density interfaces in the ocean interior tend to be an amplified mirror image of the shape of the sea surface. The process that leads to this structure is sometimes referred to as *baroclinic compensation.*

Baroclinic compensation implies that the horizontal pressure-gradient force becomes weaker with depth and so do the associated geostrophic flows. Consequently, large-scale wind-driven geostrophic flows tend to becomes vanishingly small below depths of 1500–3000 m. Whereas the sea level can approach an equilibrium state, as described by the Sverdrup relation (Eqs. 6.47 and 6.48), density interfaces in the ocean interior reach an equilibrium only in the presence of additional ageostrophic effects such as provided by lateral momentum diffusion.

6.10.2 Aim

The real ocean has a density stratification. Excess of solar heating at tropical and subtropical latitudes produces a warm surface layer that is separated from the cold abyss by a temperature transition zone, called the *permanent thermocline*. Dependent on location, the permanent thermocline extends to depths of 500–2000 m. The simplest model of this stratification is a two-layer ocean in which the density interface represents the thermocline. The aim of this exercise to explore the wind-driven circulation of the ocean in a fluid of two superimposed layers of different densities.

6.10.3 Task Description

The exercise is a repeat of Exercise 18 (Scenario 3), but with consideration of a two-layer ocean. The top layer has an initial thickness of 200 m and a density of 1025 kg/m^3. The bottom layer has an initial thickness of 800 m and a density of 1030 kg/m^3. The nonlinear terms are enabled. Horizontal eddy viscosity is set to 500 m^2/s and the bottom-friction coefficient is chosen as $r = 0.001$ m/s. The total simulation time is 100 days with data outputs at every 2.5 days. The time step is set to $\Delta t = 20$ s.

6.10.4 Results

The effect of bottom friction results in the anticipated overall downward displacement of the density interface owing to baroclinic compensation (Fig. 6.19). Maximum speeds of 75 cm/s are created in the top layer in the western boundary current. Speeds in the bottom layer rarely exceed 3 cm/s during the simulation. The maximum vertical displacement of the density interface is 85 m. A cyclonic eddy forms in the eastward return path of the western boundary current being accompanied by an upward displacement of the density interface.

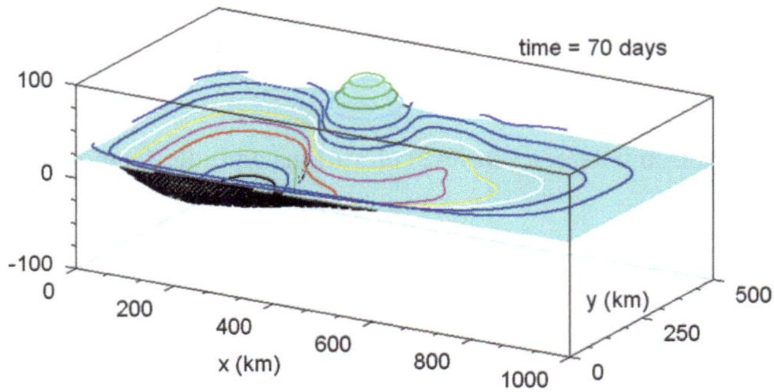

Fig. 6.19 Exercise 19. Shape of the density interface (thermocline) after 70 days of simulation

6.10.5 Sample Code and Scilab Animation Script

This exercise employs an extended version of the multi-layer shallow-water model, described in Sect. 4.5. The folder "Exercise 19" of the CD-ROM contains a full version of this code including wind forcing, bottom friction, the nonlinear terms, lateral momentum diffusion, lateral friction and the beta-plane approximation. The file "info.txt" gives additional information.

6.10.6 Additional Exercise for the Reader

Include a mid-ocean ridge in the bathymetry and study how the wind-driven circulation of a two-layer ocean responds to this. Experiment with different ridge heights and widths. Does the ridge have an impact on the shape of the density interface?

6.11 The Reduced-Gravity Concept

6.11.1 Background

Perfect baroclinic compensation in the ocean implies the absence of horizontal pressure gradients below a certain depth level, called the *level-of-no-motion*. Under this assumption, sea-level elevations can be derived from vertical displacements of density interfaces.

6.11.2 The Rigid-lid Approximation

The essence of the *rigid-lid approximation* is the assumption that the density surface of the bottom-nearest model layer always adjusts such that there is no flow in this layer after each finite time step. In a two-layer ocean, for instance, this assumption implies that:

$$P_2 = 0 = \rho_1\, g\, \eta_1 + (\rho_2 - \rho_1)\, g\, \eta_2 \tag{6.54}$$

leading to the relation between sea-level elevations and interface displacements:

$$\eta_1 = -\frac{\rho_2 - \rho_1}{\rho_1}\eta_2 \tag{6.55}$$

Oceanographers use this relation to estimate slopes of the sea level, driving the surface geostrophic flow, from the slope of the permanent thermocline. With the settings of Exercise 19, for instance, the latter relation suggests an interface displacement of 20.5 m per 10 cm of sea-level elevation. The reader is encouraged to verify this against the simulation results.

Using (6.55), the reduced-gravity version of the shallow-water wave equations for a one-dimensional channel is given by:

$$\frac{\partial u_1}{\partial t} = +g'\frac{\partial \eta_2}{\partial x} \tag{6.56}$$

$$\frac{\partial \eta_2}{\partial t} = +\frac{\rho_2}{\rho_1}\frac{\partial\,(u_1\, h_1)}{\partial x} \tag{6.57}$$

where reduced gravity is defined by $g' = (\rho_2 - \rho_1)/\rho_1\, g$. Equation (6.57) can be derived from volume conservation of the upper layer; that is,

$$\frac{\partial h_1}{\partial t} = \frac{\partial(\eta_1 - \eta_2)}{\partial t} = -\frac{\partial\,(u_1\, h_1)}{\partial x} \tag{6.58}$$

with insertion of (6.55). Under the assumption that interface displacements attain much larger amplitudes than the sea surface, $|\eta_2| >> |\eta_1|$, the thickness of the upper layer can be approximated as $h_1 \approx h_{1,o} - \eta_2$ with $h_{1,o}$ being the undisturbed thickness of the surface layer (this approximation justifies the term "rigid-lid approximation"), and Eqs. (6.56) and (6.57) can be written as:

$$\frac{\partial u_1}{\partial t} = -g'\,\frac{\partial h_1}{\partial x} \tag{6.59}$$

$$\frac{\partial h_1}{\partial t} = -\frac{\partial (u_1 h_1)}{\partial x} \tag{6.60}$$

where the Boussinesq approximation, $\rho_2/\rho_1 \approx 1$, has also been applied. These equations are formally identical to those governing surface gravity waves (Eqs. 4.12 and 4.13). Under the assumption that interface displacements are small compared with the thickness of the top layer, it can be shown that the wave solution of the two-layer reduced-gravity model are interfacial waves of a phase speed of (e.g., Gill, 1982):

$$c_{iw} = \sqrt{g' h_{1,o}} \tag{6.61}$$

which is much smaller than the phase speed of long surface gravity waves. Note that this is the same result that (4.27) gives for $h_2 >> h_1$. Obviously, application of the reduced-gravity concept to the two-layer shallow-water wave equations has filtered away fast propagating surface gravity waves. Since the phase speed of internal waves is much smaller compared with that of surface gravity waves, the reduced-gravity model allows for much longer time steps than the full two-layer model, which makes this model attractive for studies of internal-wave propagation. The CFL stability criterion for the resultant finite-difference wave equations is:

$$\Delta t \le \Delta x / c_{iw} \tag{6.62}$$

Derivation of the reduced-gravity equations for more than two layers remains for the reader. FORTRAN codes of reduced-gravity layer models are not included in this book.

6.12 Geostrophic Adjustment of a Density Front

6.12.1 Background

Density fronts are narrow zones in the ocean across which the density of seawater changes rapidly. Almost uniform densities are found on either side of the frontal zone. A variety of processes create density fronts, such a freshwater runoff from rivers into the ocean. Once a density contrast is established, adjustment toward a geostrophic steady-state circulation follows. To explore this *geostrophic adjustment*, first described by Rossby (1938), we consider the initial situation of a cylindrical patch of low-density surface water surrounded by a denser ocean of uniform density (Fig. 6.20).

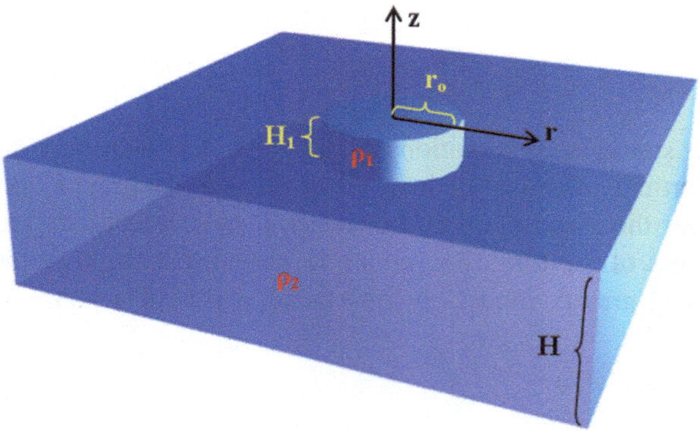

Fig. 6.20 Illustration for the description of the geostrophic adjustment process

6.12.2 How Does It Work?

Initially, lateral pressure gradients associated with the slope of the density interface creates bottom flow directed radially inward (Fig. 6.21). Consequently, the sea level will start to rise in the centre of this patch owing to the resultant flow convergence in the bottom layer. Pressure-gradients owing to a tilted sea level, on the other hand, creates surface flow directed radially outward. On time scales of the inertial period and beyond, the Coriolis force comes into play and deflects these flows. The initial radial spreading turns into circular geostrophic motion. Water-column squeezing produces a geostrophic eddy of cyclonic vorticity in the low-density surface patch. Water-column stretching produces a clockwise rotating eddy in the ocean underneath.

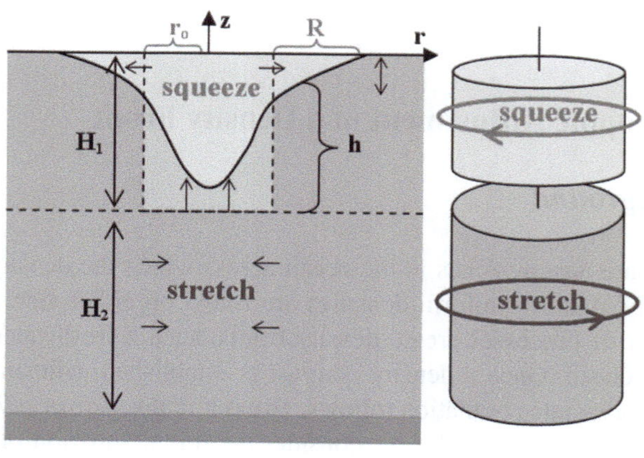

Fig. 6.21 Illustration of the geostrophic adjustment process in a 2-layer ocean and resultant geostrophic eddies for the northern hemisphere

6.12.3 Theory

The geostrophic balance governs the steady-state dynamics of the geostrophic adjustment problem. The corresponding momentum equations in cylinder coordinates read:

$$f v_1 = g \frac{\partial \eta}{\partial r} \tag{6.63}$$

$$f v_2 = g \frac{\rho_1}{\rho_2} \frac{\partial \eta}{\partial r} - g' \frac{\partial h}{\partial r} \tag{6.64}$$

where the index 1 refers to the top layer, the index 2 refers to the bottom layer, f is the Coriolis parameter, r is the radial coordinate, v is the speed of the frontal flow, η_1 is sea-surface elevation, h is the interface displacement with reference to the initial depth level H_1 (see Fig. 6.21), and reduced gravity is $g' = (\rho_2 - \rho_1)/\rho_2 \, g$. Positive speeds correspond to counterclockwise flow.

With use of the reduced-gravity concept, which implies vanishing flow in the bottom layer, and the Boussinesq approximation ($\rho_1/\rho_2 \approx 1$), the momentum equation for the surface layer can be written as:

$$f v_1 = -g' \frac{\partial h}{\partial r} \tag{6.65}$$

The reduced-gravity concept is valid if the bottom layer is much thicker than the surface layer ($H_2 >> H_1$). Potential vorticity is conserved during the geostrophic adjustment process. Consequently, the initial and final states have the same potential vorticity, which can be expressed as:

$$\frac{f + \xi_1}{H_1 - h} = \frac{f}{H_1}$$

where relative vorticity in the top layer is defined by $\xi_1 = \partial v_1 / \partial r$. This equation can be rewritten as:

$$\xi_1 = -\frac{h}{H_1} f \tag{6.66}$$

Equations (6.65) and (6.66) can be combined to yield an equation in the single variable h; that is,

$$\frac{\partial^2 h}{\partial r^2} = \frac{h}{R^2} \tag{6.67}$$

where the *internal Rossby radius of deformation* is given by:

$$R = \frac{\sqrt{g' H_1}}{|f|} \tag{6.68}$$

The solution of (6.67) is (e.g., Cushman-Roisin, 1994):

$$h(r) = H_1 \exp\left(\frac{r - r_o - R}{R}\right) \tag{6.69}$$

where r_o is the initial radius of the low-density patch with $r_o >> R$. The internal Rossby radius of deformation gives an estimate of the frontal width. The solution for geostrophic flow in the surface layer (northern hemisphere) follows from (6.65) and is given by:

$$v_1(r) = -\sqrt{g' H_1} \exp\left(\frac{r - r_o - R}{R}\right) \tag{6.70}$$

The flow direction is reversed for the southern hemisphere. In this analytical solution, frontal flows attain the swiftest speeds at the location where the density interface outcrops at the sea surface, whereas there are no flows just outside this front. Such a discontinuity cannot exist in the real world. Instead of this, lateral friction produces a transition zone across the front and frontal flow speeds tend to be smaller than predicted by theory. Interestingly, although the steady-state frontal flow is purely geostrophic, its magnitude is independent of the Coriolis parameter. It should also be noted that the maximum frontal speed equals the phase speed of long internal waves. Typical oceanic values of v_1 and R, respectively, are 0.1–0.5 m/s and 1–5 km. Geostrophic adjustment can be expected to occur on a time scale exceeding several inertial periods.

An isolated layer of dense water on the seafloor also becomes subject to the geostrophic adjustment process (Fig. 6.22). Under the assumptions that there are no flows outside this layer and absence of frictional effects, the steady-state momentum equation for the bottom layer can be written as:

$$f v_2 = -g' \frac{\partial h}{\partial r}$$

where h is the downward displacement of the density interface with reference to the initial thickness H_2. Conservation of potential vorticity can be expressed as:

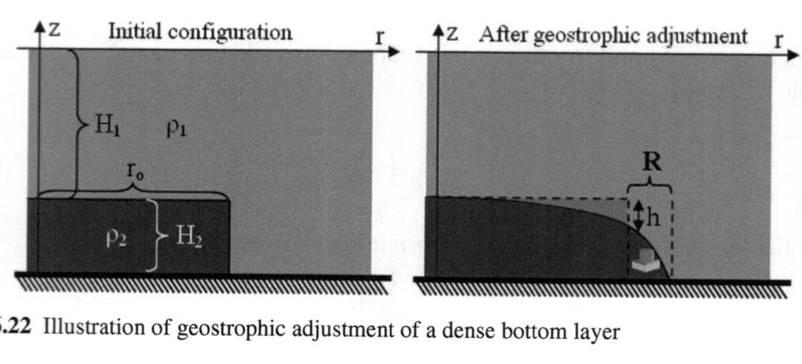

Fig. 6.22 Illustration of geostrophic adjustment of a dense bottom layer

$$\xi_2 = -\frac{h}{H_2} f$$

where relative vorticity in the bottom layer is defined by $\xi = \partial v_2/\partial r$. The latter two equations can be combined to yield an equation of the same form as (6.67). The solution for the northern hemisphere is:

$$h(r) = H_2 \exp\left(\frac{r - r_o - R}{R}\right)$$
$$v_2(r) = -\sqrt{g' H_2} \exp\left(\frac{r - r_o - R}{R}\right)$$

where r_o is the initial radius, and the internal Rossby radius of deformation is now given by:

$$R = \frac{\sqrt{g' H_2}}{|f|}$$

The geostrophic adjustment process is mathematically more difficult to describe for situations in which both layers have comparable thicknesses and therefore not included in this book. Generally, geostrophic flows in the top and bottom layers are opposite to each other and the ratio of speeds depends on the initial thicknesses of layers involved. The frontal width is again given by the internal deformation radius:

$$R = \frac{\sqrt{g' H^*}}{|f|} \tag{6.71}$$

where the "equivalent" thickness is given by $H^* = H_1 H_2/(H_1 + H_2)$. Note that the numerator in the latter relation is the phase speed of long internal gravity waves for a two-layer fluid (4.27).

6.13 Exercise 20: Geostrophic Adjustment

6.13.1 Aim

The aim of this exercise is to explore the geostrophic adjustment process for an ocean of two superimposed layers of different densities.

6.13.2 Task Description

The model domain is 200 m deep and has equal sidelengths of 50 km. The surface patch of lower-density water has a density of 1027 km m^{-3}, a thickness of 100 m and a radius of 15 km (see Fig. 6.20). The ambient ocean has a density of 1028 km m^{-3}. The Coriolis parameter is set to $f = 1 \times 10^{-4}$ s^{-1}. The task is to employ the two-layer shallow-water model to simulate the geostrophic adjustment process over 5 days with horizontal grid spacings of $\Delta x = \Delta y = 1$ km. The time step is set to $\Delta t = 2$ s. After (6.68), the expected width of the frontal zone can be estimated at about 10 km. The horizontal grid spacing chosen just resolves this scale.

This exercise ignores wind-stress-forcing, the nonlinear terms, and horizontal and vertical friction. All lateral boundaries are kept open using zero-gradient conditions for all variables. To avoid the appearance of unwanted gravity waves and inertial oscillations, the density anomaly in the surface layer is slowly linearly adjusted from zero to its final value over the first 2 days of the simulation. Non-buoyant Lagrangian floats are included to visualise the resultant flow paths.

6.13.3 Results

Figure 6.23 reveals that, owing to geostrophic adjustment, the low-density surface patch still exists after 5 days of simulation. In fact, in the absence of friction, the lifetime of such an eddy is unlimited. The maximum thickness of this patch decreases from the initial 100 m to a steady-state value of about 50 m within the first 2 days of simulation. This decrease in thickness is associated with the generation of a sequence of internal waves propagating energy radially outward during the initial adjustment phase.

In agreement with the principle of conservation of potential vorticity, an anticyclonic (clockwise-rotating) geostrophic eddy establishes in the surface layer with frontal flow speeds of 33 cm/s (Fig. 6.24a). A cyclonic geostrophic eddy of a maximum speed of 10 cm/s establishes in the bottom layer (Fig. 6.24b). Note that the surface and bottom eddies rotate in opposite directions in agreement with expectations (see Fig. 6.21).

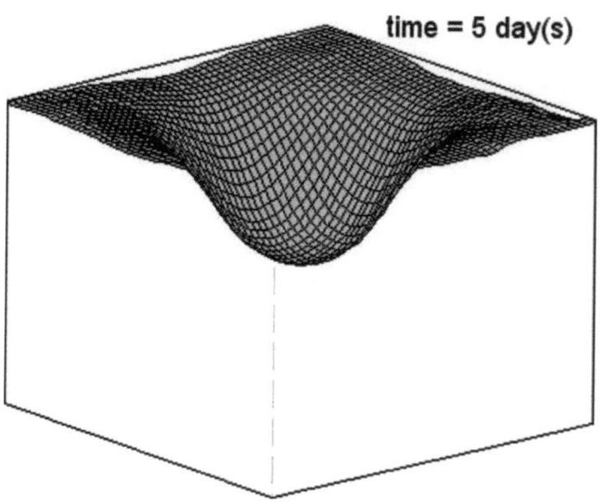

Fig. 6.23 Exercise 20. Shape of the density interface after 5 days of simulation. The steady-state maximum thickness of the upper layer is 50 m

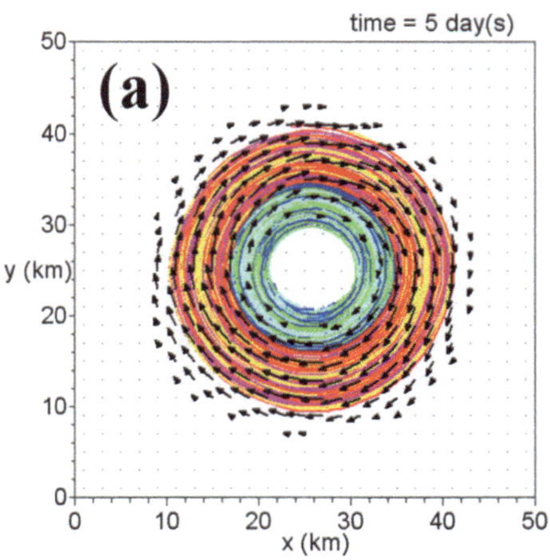

Fig. 6.24 Exercise 20. Flow field (*arrows*) in the (**a**) surface and (**b**) bottom layers after 5 days of simulation. Maximum flow speed in the surface layer is 33 cm/s and in the bottom layer this is 10 cm/s. Velocity vectors are averaged over 2×2 grid cells. Curves are trajectories of 400 Lagrangian floats predicted over the last 3 days of simulation

With reference to the equilibrium state with $H_1 \approx 50$ m in the centre of the eddy, Eqs. (6.68) and (6.70) give a frontal width $R \approx 7$ km and a frontal speed of 70 cm/s, which is different from the model predictions. The analytical solutions, however, assume vanishing flow in the bottom layer, which does not reflect the situation considered in this exercise. It should be highlighted that the flooding algorithm does not lead to significant problems in this exercise.

6.13.4 Sample Code and Animation Script

The folder "Exercise 20" of the CD-ROM contains the computer codes for this exercise. The file "info.txt" gives additional information.

6.13.5 Additional Exercise for the Reader

Repeat this exercise for the southern hemisphere situation with $f = -1 \times 10^{-4}\,\mathrm{s}^{-1}$. Does the model prediction agree with your expectations?

6.14 Baroclinic Instability

6.14.1 Brief Description

In addition to the barotropic instability mechanism (see Sect. 6.6), quasi-geostrophic flow can become subject to another form of instability, called *baroclinic instability*, first analytically described by Eady (1949). The source of this instability is a depth-variation of horizontal geostrophic flow in a stratified ocean associated with tilted density interfaces. Disturbances in such a baroclinic geostrophic flow leads to water-column stretching and squeezing at different locations of the disturbance. This generates self-enforcing patterns of relative vorticity and disturbances grow in time. It can be shown that perturbations of a wavelength of about fourfold the internal deformation radius have the greatest initial growth rate. Cushman-Roisin (1994) presents the theory describing this instability process.

The baroclinic instability mechanism is the origin of the large mid-latitude cyclones and anticyclones that make our weather so variable and creates eddies in the ocean. Geostrophic frontal flows, such as those inherent with western boundary currents, exhibit the steepest slopes of density interfaces and are therefore subtle to the baroclinic instability mechanism. In the real situation, frontal flows can become unstable to both barotropic instability owing to lateral current shear and baroclinic instability owing to vertical current shear. It is often difficult to tell which mechanism was the major cause of instability development.

Is has been long thought that flows in the deep ocean are generally weak. This, however, is not always the case. Frontal instabilities of western boundary currents, for instance, trigger mesoscale eddies that produce swift flow in the abyssal ocean, referred to as *benthic storms*, first observed by Rowe and Menzies (1968). Benthic storms can attain speeds >50 cm/s on timescales of 20 days and are capable of eroding sediment from the abyssal seafloor. Hence, the hypothesis that the deep oceans are quiescent, as suggested by the baroclinic compensation process, is not valid in these frontal regions.

The theory behind the baroclinic instability process is complex and therefore not included in this book. Nevertheless, we can employ the multi-layer model, developed in previous exercises, to investigate this process.

6.15 Exercise 21: Frontal Instability

6.15.1 Aim

The aim of this exercise is to explore frontal instabilities of quasi-geostrophic flows in an ocean of two superimposed layers of different densities.

6.15.2 Task Description

The model domain has a length of 200 km, a width of 100 km, and a depth of 500 m with closed boundaries in the north and in the south. Cyclic boundary conditions are used at the western and eastern boundaries of the model domain. Lateral grid spacings are set to $\Delta x = \Delta y = 2$ km. The ocean is approximated by a two-layer system. Density of the top layer is chosen at $\rho_1 = 1027.25$ kg m^{-3}. Density of the bottom layer is set to $\rho_2 = 1028$ kg m^{-3}.

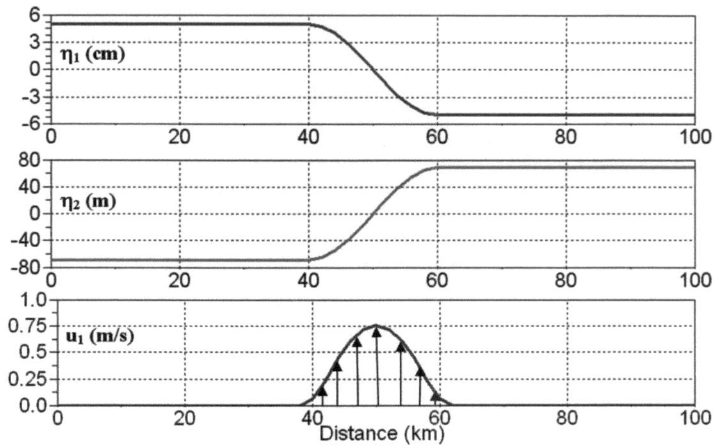

Fig. 6.25 Exercise 21. Initial distributions across the front of (*top panel*) sea-level elevation (cm), (*middle panel*) interface displacement, and (*bottom panel*) and geostrophic frontal flow

The frontal flow is initiated by prescription of sea-level variations in the y-direction (Fig. 6.25) using a sine function for the frontal transition zone. Random noise of an amplitude of 5 mm is added to sea-level elevations to facilitate the inital growth of disturbances.

The initial elevation of the density interface is calculated from (6.54), which implies that, initially, the bottom layer is at rest. In this exercise, the depth of the density interface varies by ± 73 m across the front. The Coriolis parameter is set to $f = 1 \times 10^{-4}$ s^{-1}. The internal deformation radius associated with the initial configuration is about 7.5 km, so that we expect disturbances to grow on a lengthscale of 30 km.

Initial speeds of the upper-layer frontal flow are calculated from the geostrophic balance (Eq. 6.13). The frontal flow in the upper layer attains maximum speeds of 80 cm/s and a width of the frontal zone is 20 km. The total simulation time is 20 days with data outputs at every 6 h. The time step is set to $\Delta t = 2$ s.

Eulerian tracer concentration is added to visualise cross-frontal disturbances. To this end, tracer concentration of unity is added initially to the southern half of the channel, whereas the other hand is initialised with zero values. The nonlinear terms are essential in the instability process and need to be included in this exercise. Bottom friction, lateral momentum diffusion and lateral friction can be ignored to first-order approximation.

6.15.3 Results

Flow disturbances start to grow on wavelengths of 40 km after 8 days of simulation and manifest themselves in disturbances in the density interface (Fig. 6.26). The wavelength of predominant disturbances of $\tilde{2}0$ km agrees with theory. As a result

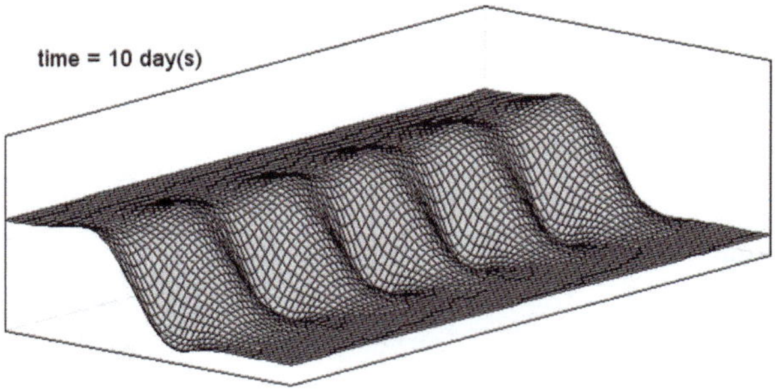

Fig. 6.26 Exercise 21. Shape of the density interface after 10 days of simulation

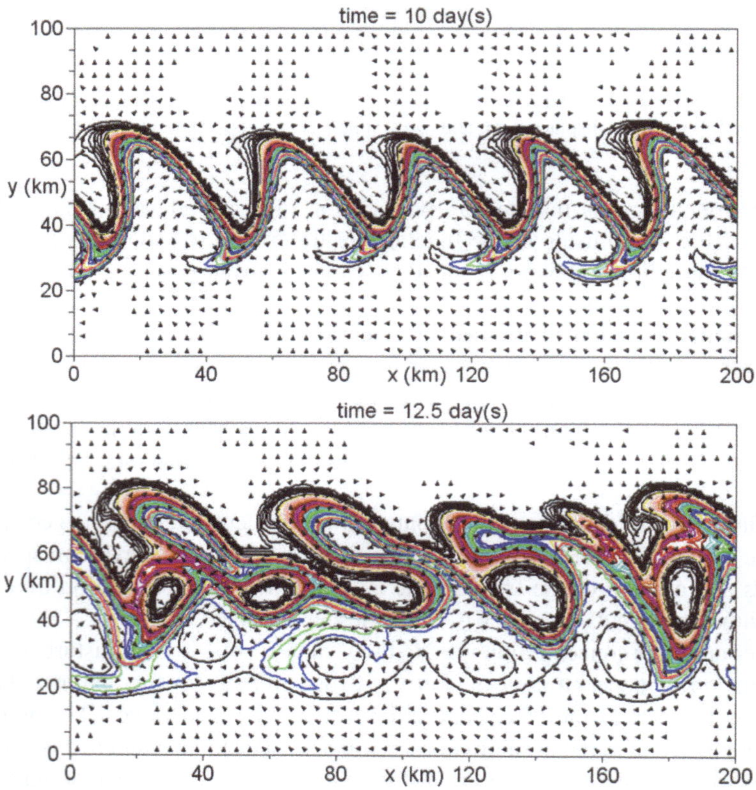

Fig. 6.27 Exercise 21. Currents (*arrows*, averaged over 2×2 grid cells) and Eulerian tracer concentration (contours) in the top layer after 10 and 12.5 days of simulation. Cross-frontal flow attains speeds of 50 cm/s

of this instability, the front starts to meander forming alternating zones of positive and negative relative vorticity (Fig. 6.27). Baroclinic eddies appear soon after this mixing whereby cyclonic eddies remain centered along the axis of the front while anticyclonic eddies are moved away from the frontal axis. The diameter of eddies

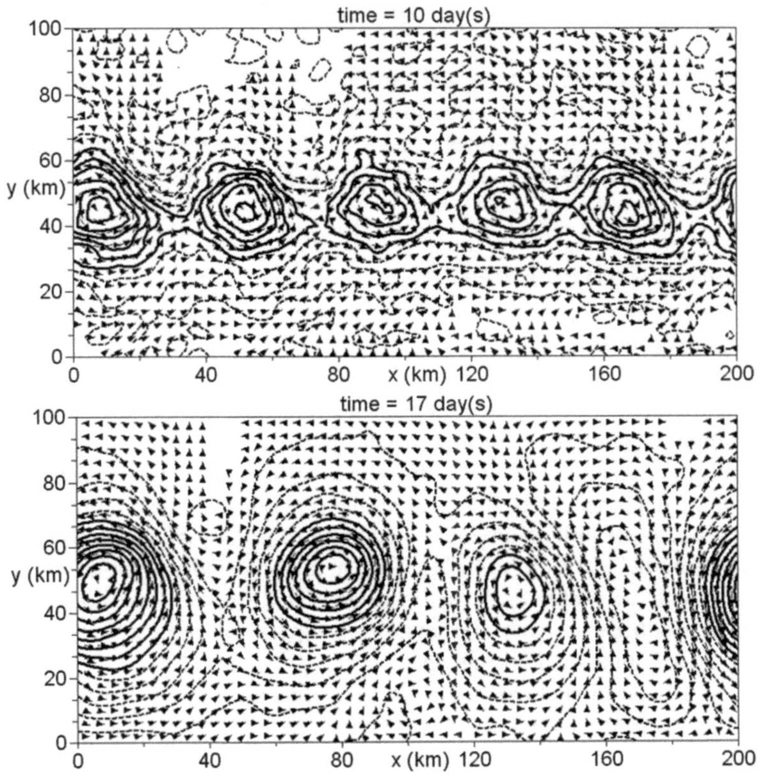

Fig. 6.28 Exercise 21. Currents (*arrows*, averaged over 2×2 grid cells) and pressure anomalies (solid contours emphasise low pressure centres) in the bottom layer after 10 and 17 days of simulation. Deep cyclones attain maximum flow speeds of 20 cm/s

is 10–20 km. Later in the simulation (not shown), eddies interact with each other and produce vigorous lateral mixing in the entire model domain. Surface eddies are associated with swift currents of speeds exceeding 50 cm/s.

Bottom currents of noticeable speed (benthic storms) start to develop from day 10 of the simulation onward and approach values of >20 cm/s by the end of the simulation (Fig. 6.28). Interestingly, the structure of disturbances in the bottom layer is different from those in the upper layer. Here it is exclusively the cyclonic eddies that trigger the growth of instabilities. This process leading to the preferred creation of cyclonic eddies is called *cyclogenesis*.

6.15.4 Sample Code and Animation Script

The folder "Exercise 21" of the CD-ROM contains the computer codes for this exercise. The "info.txt" file gives more information.

6.15.5 Additional Exercise for the Reader

Repeat this exercise for an ocean uniform in density ($\rho_1 = \rho_2 = 1028\,\text{kg m}^{-3}$) to explore whether the barotropic instability process alone can produce a similar form of frontal instability. The two-layer version of the shallow-water equations can be adopted for this task, but the reader should avoid division by zero...

6.16 Density-Driven Flows

6.16.1 Background

Density-driven flows are bodies of dense water that cascade downward on the continental slope to a depth where they meet ambient water of the same density. At this equilibrium density level, these flows tend to flow along topographic contours of the continental slope. Detachment from the seafloor and injection into the ambient ocean is also possible. Numerical modelling of gravity plumes started with the "streamtube model" of Smith (1975). This model considers a laterally integrated streamtube with a variable cross-sectional area and, under the assumption of stationarity, it produces the path and laterally averaged properties (density contrast, velocity) of the plume on a given slope.

With an advanced method describing the dynamics of such *reduced-gravity plumes*, Jungclaus and Backhaus (1994) employed the shallow-water equations for a two-layer fluid with an upper layer at rest. Without exchange of fluid across the "skin" of the plume, the dynamic governing this model can be written as:

$$\frac{\partial u_2}{\partial t} + u_2\frac{\partial u_2}{\partial x} + v_2\frac{\partial u_2}{\partial y} - f v_2 = -g'\frac{\partial \eta_2}{\partial x} - \frac{\tau_x^{\text{bot}}}{h_2\rho_2}$$
$$\frac{\partial v_2}{\partial t} + u_2\frac{\partial v_2}{\partial x} + v_2\frac{\partial v_2}{\partial y} + f u_2 = -g'\frac{\partial \eta_2}{\partial y} - \frac{\tau_y^{\text{bot}}}{h_2\rho_2} \tag{6.72}$$
$$\frac{\partial \eta_2}{\partial t} + \frac{\partial (u_2 h_2)}{\partial x} + \frac{\partial (v_2 h_2)}{\partial y} = 0$$

where $g' = (\rho_2 - \rho_1)/\rho_2\, g$ is reduced gravity, and $(\tau_x^{\text{bot}}, \tau_y^{\text{bot}})$ represents the frictional stress at the seafloor. These equations are known as the *reduced-gravity plume model.* Interface displacements η_2 are defined with respect to a certain reference level. When formulated in finite differences, the CFL stability criterion in such a model is given by:

$$\Delta t \le \frac{\min(\Delta x, \Delta y)}{\sqrt{2g' h_{\text{max}}}}$$

where h_{max} is maximum plume thickness.

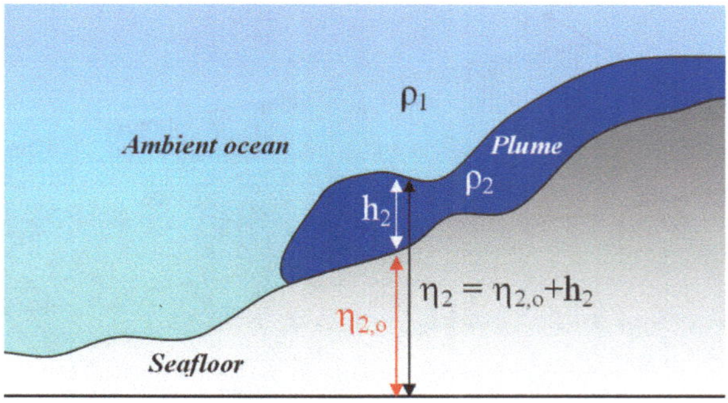

Fig. 6.29 Definition of interface displacement and layer thickness for the reduced-gravity plume model

For simplicity, we assume that the plume is denser than any ambient water, so that the deepest part of the model domain can be chosen as reference level. The tilt of the surface of the plume is then calculated with reference to this level (Fig. 6.29). This implies that with initial absence of a plume, interface displacements $\eta_{2,o}$ have to follow the shape of the bathymetry. This is similar to treatment of sloping coasts in the flooding algorithm (see Sect. 4.4).

6.17 Exercise 22: Reduced-Gravity Plumes

6.17.1 Aim

The aim of this exercise is to explore the dynamics inherent with the descent of a reduced-gravity plume on a sloping seafloor.

6.17.2 Task Description

We consider a model domain of 200 km in length and 100 km in width (Fig. 6.30), resolved by lateral grid spacings of $\Delta x = \Delta y = 2$ km. The seafloor has a mild uniform upward slope of 200 m per 100 km in the y-direction. Since the surface ocean is at rest in this reduced-gravity plume model, the total water depth is irrelevant here. Frictional stresses at the seafloor are described by a quadratic bottom-friction law.

An artificial coastline is placed along the shallow side of the model domain, except for a narrow opening of 6 km in width used as a source for the reduced-gravity plume to enter the model domain. The initial thickness of the plume is set to 100 m in this opening. The density excess of the plume is set to 0.5 kg m^{-3}. Density of ambient water is $\rho_1 = 1027$ kg m^{-3}. Zero-gradient conditions are used at open boundaries for all variables.

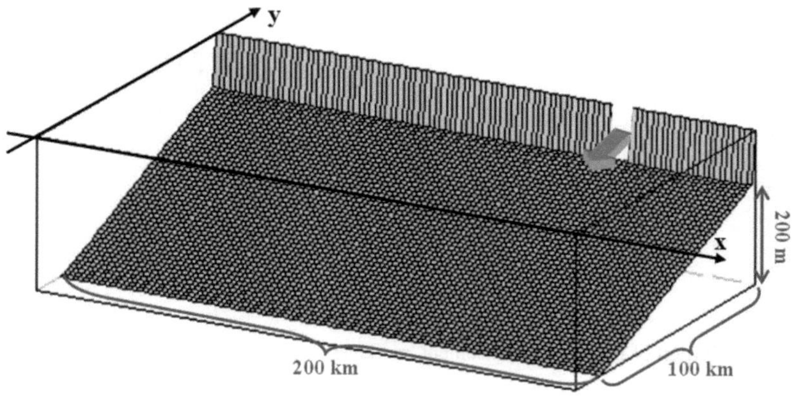

Fig. 6.30 Bathymetry for Exercise 22 (Scenario 2)

Two different scenarios are considered. The Coriolis force is ignored in the first scenario. Case studies consider variations of bottom drag coefficient. The total simulation time is one day with data outputs at every hour. Because we expect a symmetric shape of the plume, the forcing region is placed in the centre of the otherwise closed boundary that cuts along shallower regions of the model domain.

The second scenario includes the Coriolis force with $f = +1 \times 10^{-4}\ \mathrm{s}^{-1}$ (northern hemisphere). Again, case studies consider variations of values of the bottom-drag coefficient. The total simulation time is 5 days with one-hourly data outputs. In anticipation of rotational effects imposed by the Coriolis force, the forcing region is moved some distance. This is why the forcing region has been moved some distance upstream, as is shown in Fig. 6.30. The time step is set to $\Delta t = 6\ \mathrm{s}$ in all experiments.

6.17.3 Write a New Simulation Code?

There is no need to formulate a new FORTRAN simulation code for this exercise. Instead, the two-layer of this, the two-layer version of the shallow-water equations, used in Exercises 20 and 21, can be applied with the constraint that the surface layer is at rest.

6.17.4 Results

As anticipated, the forcing applied creates a gravity current moving denser water away from the source. First, we consider the situation without the Coriolis force (Scenario 1). On an even seafloor, the spreading of dense water would be radially symmetric. On the other hand, a sloping seafloor supports a net downslope pressure-gradient force, so that, in addition to radial spreading, the plume moves downslope

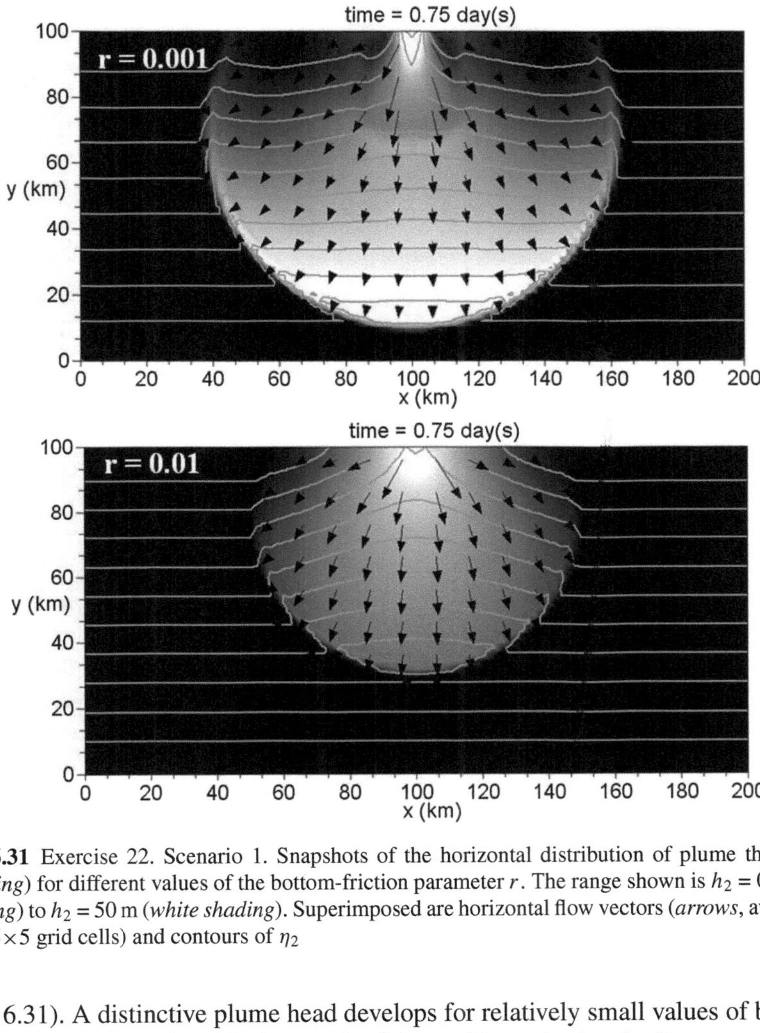

Fig. 6.31 Exercise 22. Scenario 1. Snapshots of the horizontal distribution of plume thickness (*shading*) for different values of the bottom-friction parameter r. The range shown is $h_2 = 0$ (*black shading*) to $h_2 = 50$ m (*white shading*). Superimposed are horizontal flow vectors (*arrows*, averaged over 5×5 grid cells) and contours of η_2

(Fig. 6.31). A distinctive plume head develops for relatively small values of bottom friction. This plume head is the result of the rapid gravitational adjustment taking place during the initial phase of the simulation. Increased values of the bottom-friction parameter lead to both disappearance of the plume head and overall weaker flows.

The dynamical behaviour of reduced-gravity plumes is turned "upside down" with inclusion of the Coriolis force. With relatively low levels of bottom friction, the plume rather follows bathymetric contours with only little tendency of downslope motion (Fig. 6.32). With presence of the Coriolis force, it is increased levels of bottom friction that induce an enhanced angle of downslope motion. Consideration of simple steady-state dynamical balances between the reduced-gravity force, the Coriolis force and the frictional force explains this interesting feature (Fig. 6.33).

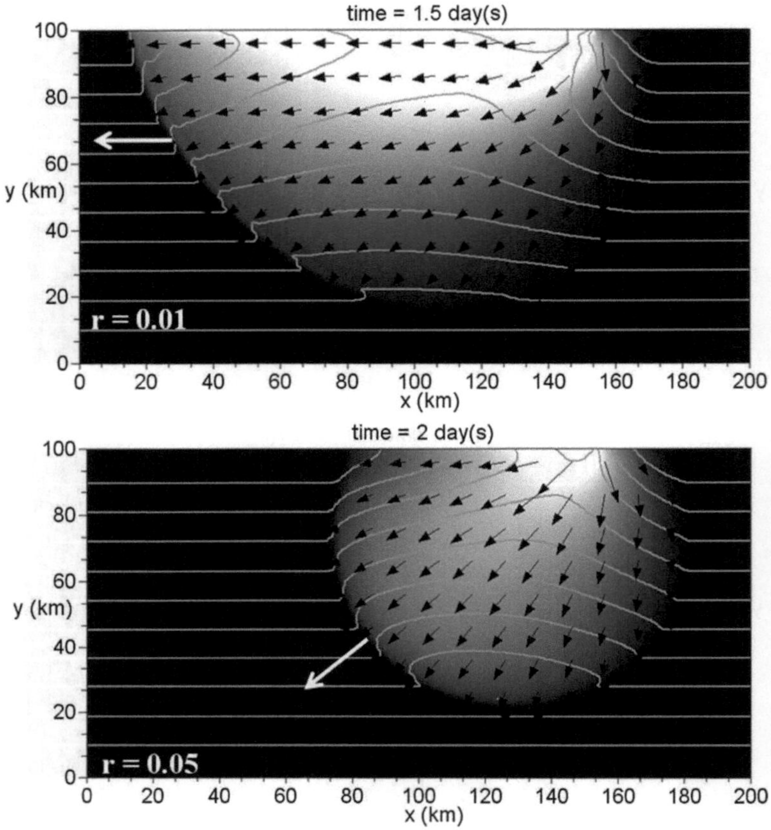

Fig. 6.32 Same as Fig. 6.31, but for Scenario 2

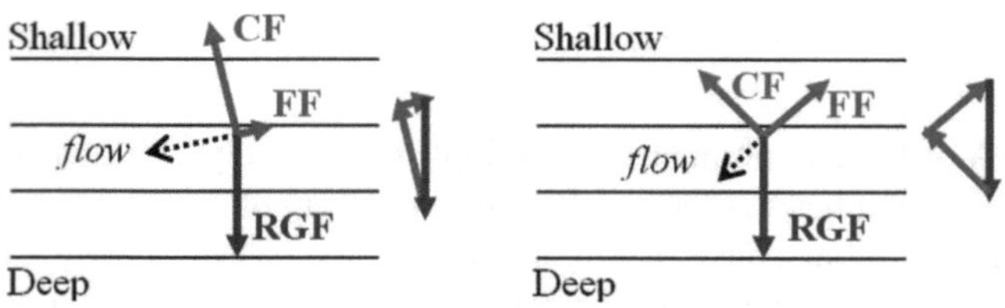

Fig. 6.33 Steady-state force balances of reduced-gravity plumes for various levels of bottom friction on the northern hemisphere. The reduced-gravity force (RGF) acts downward on the sloping seafloor. The Coriolis force (CF) acts at a right angle with respect to the flow direction. Bottom friction (BF) acts opposite to the flow direction

6.17.5 Sample Code and Animation Script

The folder "Exercise 22" of the CD-ROM contains the computer codes for this exercise. See the file "info.txt" for more information.

6.17.6 Additional Exercise for the Reader

Add a topographic obstacle such as a seamount or a seafloor depression to the bathymetry and explore how reduced-gravity plumes deal with irregular bathymetry.

6.18 Technical Information

This book has been written in LaTeX using TeXnicCenter, downloadable at

http://www.toolscenter.org/

in conjunction with MikTeX (Version 2.5) – a LaTeX implementation for the Windows platform, which can be downloaded at:

http://miktex.org/

Most graphs of this book were created with SciLab. GIMP has been used for the manipulation of some images. GIMP is a cost-free alternative to commercial graphical manipulation programs such as Adobe Photoshop. This software is freely downloadable at:

http://www.gimp.org/

Most sketches were made in Microsoft Word. Figure 3.21 was produced with BLENDER (Version 2.43) – a three-dimensional animation suite (and game engine) available at:

http://www.blender.org/

Bibliography

Arakawa, A., and Lamb, V., 1977: Computational design of the basic dynamical processes of the ucla general circulation model, in Methods in Computational Physics, vol. 17, pp. 174–267. Academic Press London.

Boussinesq, J., 1903: Théorie Analytique de la Chaleur. Vol. 2. Gautier Villars, Paris.

Brunt, D., 1927: The period of simple vertical oscillations in the atmosphere. Q. J. R. Met. Soc., 53, 30.

Coriolis, G. G., 1835: Mémoire sur les équations du mouvement relatif des systèmes de corps. Journal de l'école Polytechnique, 15, 142–154.

Courant, R., Friedrichs, K., and Lewy, H., 1928: Über die partiellen Differenzengleichungen der mathematischen Physik, Mathematische Annalen, Vol. 100, No. 1, 32–74.

Cushman-Roisin, B., 1994: Introduction to Geophysical Fluid Dynamics, Prentice Hall, Englewood, Cliffs 320 pp.

Eady, E., 1949: Long waves and cyclone waves. Tellus, 1, 33–52.

Ekman, V. W., 1905: On the influence of the Earth's rotation on ocean-currents. Arkiv for Matematik, Astronomi och Fysik, 2 (11), 1–52.

Euler, L., 1736: Mechanica sive motus scientia analytice exposita, St. Petersburg.

Fringer, O. B., Armfield, S. W., and Street, R. L., 2005: Reducing numerical diffusion in interfacial gravity wave simulations, Int. J. Numer. Methods Fluids, 49 (3), 301–329, doi: 10.1002/fld.992.

Froude, W., 1874: On useful displacement as limited by weight of structure and of propulsive power. Trans. Inst. Naval Architects, 15, 148–155.

Gill, A. E., 1982: Atmosphere-Ocean Dynamics, Academic Press, London.

Jungclaus, J. H., and Backhaus, J. O., 1994: Application of a transient reduced gravity plume model to the Denmark Strait Overflow. J. Geophys. Res., 99 (C6), 12375–12396.

Kuo, H. L., 1949: Dynamic instability of two-dimensional non-divergent flow in a. barotropic atmosphere. J. Meteor., 6, 105–122.

Lagrange, J.-L. 1788: Mécanique Analytique, Paris.

Merian, J. R. 1828: Ueber die Bewegung tropfbarer Flüssigkeiten in Gefässen. Basel, Schweighauser. Reproduced by Vonder-Mühll, K., 1886. Math. Annalen 27: 575 ff.

Munk, W. H., 1950: On the wind-driven circulation, J. Meteor., 7, 79–93.

Navier, C. L. M. H., 1822: Memoire sur les lois du mouvement des fluides, Mem. Acad. Sci. Inst. France, 6, 389–440.

Newton, I., 1687: The Principia, A new translation by I.B. Cohen and A. Whitman, University of California Press, Berkeley 1999.

Orlanski, I., 1976: A simple boundary condition for unbounded hyperbolic flows. J. Comput. Phys., 21, 252–269.

Pond, S. G., and Pickard, L., 1983: Introductory Dynamical Oceanography, 2nd edition, Pergamon Press, Oxford, 329 pp.

Prandtl, L., 1925: Bericht über Untersuchungen zur ausgebildeten Turbulenz, Z. Angew. Math. Mech., 5, 136–139.

J. Kämpf, *Ocean Modelling for Beginners*,
DOI 10.1007/978-3-642-00820-7_BM2,

Press, W.H., Flannery, B.P., Teukolsky, S.A., and Vetterling, W.T., 1989: Numerical Recipes. The Art of Scientific Computing. Cambridge University Press, Cambridge, 702 pp.

Reynolds, O., 1883: An experimental investigation of the circumstances which determine whether the motion of water shall be direct or sinous, and of the law of resistance in parallel channels. Philos. Trans. R. Soc., 174, 935–982.

Reynolds, O., 1895: On the dynamical theory of incompressible viscous fluids and the determination of the criterion. Philos. Trans. R. Soc., 186, 123–164.

Richardson, L. F., 1920: The supply of energy from and to atmospheric eddies. Pros. Roy. Soc. London, A 97, 354–373.

Rossby, C. G., 1937: The mutual adjustment of pressure and velocity distributions in certain simple current systems, I. J. Mar. Res., 1, 15.

Rossby, C. G., 1938: The mutual adjustment of pressure and velocity distribution in certain simple current systems, II. J. Mar. Res., 1, 239.

Rowe, G. T., and Menzies, R. J., 1968: Deep bottom currents off the coast of North Carolina. Deep-Sea Res., 15, 711–719.

Shapiro, R., 1970: Smoothing, filtering, and boundary effects, Rev. Geophys. Space Phys., 8, 359–387.

Smith, P. C., 1975: A streamtube model for bottom boundary currents in the oceans. Deep-Sea Res., 22, 853–873.

Sommerfeld, A., 1949: Partial differential equations. Lectures in Theoretical Physics 6. Academic Press, New York.

Stein, S., 1999: Archimedes: What Did He Do Besides Cry Eureka? Mathematical Association of America, Washington, DC, 155 pp.

Stokes, G. G., 1845: On the theories of the internal friction of fluids in motion. Trans. Cambridge Philos. Soc., 8.

Stommel, H. M., 1948: The westward intensification of wind-driven currents. Trans. Am. Geophys. Union, 99, 202–206.

Stommel, H. M., 1987: A View of the Sea. Princeton University Press, Princeton, NJ, 165 pp.

Stommel, H. M., and Moore, D. W., 1989: An Introduction to the Coriolis Force. Columbia University Press, New York, 297 pp.

Sverdrup, H. U., 1947: Wind-driven currents in a baroclinic ocean; with application to the equatorial currents of the eastern Pacific, Proc. N.A.S. 33, 318–326.

Taylor, B., 1715: Methodus incrementorum directa et inversa (Direct and Inverse Incremental Method), G. Innys, London.

Thomson, W. (Lord Kelvin), 1879: On gravitational oscillations of rotating water. Proc. Roy. Soc. Edinburgh, 10, 92–100.

Väisälä, V., 1925: Über die Wirkung der Windschwankungen auf die Pilotbeobachtungen. Soc. Sci. Fenn. Commentat. Phys. Math. 2(19), 19.

von Kármán, T., 1911: Über den Mechanismus den Widerstands, den ein bewegter Körper in einer Flüssigkeit erfährt, Gött. Nachr, Part 1: 509–517.

List of Exercises

Index

F

G

H

I

K

L

M

N

P

R

S

MIX
Papier aus verantwortungsvollen Quellen
Paper from responsible sources
FSC® C105338

If you have any concerns about our products,
you can contact us on
ProductSafety@springernature.com

In case Publisher is established outside the EU,
the EU authorized representative is:
Springer Nature Customer Service Center GmbH
Europaplatz 3, 69115 Heidelberg, Germany

Printed by Libri Plureos GmbH
in Hamburg, Germany